DEVELOPMENT OF THE BUILT ENVIRONMENT

ABOUT THE AUTHOR

DEWBERRY, headquartered in Fairfax, Virginia, is a fully integrated engineering and architecture firm operating in more than 50 locations throughout the United States. It consistently ranks among the top 45 design firms by *Engineering News-Record*, top 20 engineering/architecture firms by *Building Design + Construction*, and the top 5 engineering firms by the *Washington Business Journal*. Working in multiple federal, state and local, and commercial markets, Dewberry's services include site/civil engineering and surveying; transportation, transit, and ports and intermodal design; water, wastewater, and water resources engineering; architectural and interior design; environmental, coastal engineering and resilience services; emergency management and mitigation; full-service geospatial mapping and analysis; and alternative project delivery inclusive of design-build, public-private partnerships, and turnkey construction. The firm enjoys a reputation for quality, deep subject-matter expertise, community engagement, and putting the client first.

DEVELOPMENT OF THE BUILT ENVIRONMENT

FROM SITE ACQUISITION TO PROJECT COMPLETION

Land Development Handbook Series

Dewberry

Editor-in-Chief: Sidney O. Dewberry, PE, LS

Editor: C. Kat Grimsley, PhD cantab

New York   Chicago   San Francisco   Athens   London
Madrid   Mexico City   Milan   New Delhi
Singapore   Sydney   Toronto

Library of Congress Control Number: 2018954684

Development of the Built Environment: From Site Acquisition to Project Completion

1 2 3 4 5 6 7 8 9   LWI   23 22 21 20 19

ISBN 978-1-260-44073-7
MHID 1-260-44073-7

The pages within this book were printed on acid-free paper.

Sponsoring Editor
Lauren Poplawski

Proofreader
Manish Tiwari

Editorial Supervisor
Donna M. Martone

Indexer
Anand Shekhar

Acquisitions Coordinator
Elizabeth Houde

Production Supervisor
Pamela A. Pelton

Project Manager
Radhika Jolly,
Cenveo® Publisher Services

Composition
Cenveo Publisher Services

Art Director, Cover
Jeff Weeks

Copy Editor
Surendra Shivam

*To the hardworking Dewberry employees
who dedicate their talent, energy, and passion to building amazing places.*

CONTENTS

FOREWORD

Real estate development is akin to weaving. It is the complex intertwining of entrepreneurial vision, market/financial feasibility research, land suitability studies, site evaluation and planning, strategy, engineering, environmental and sustainability considerations, politics, legal issues, construction management, design and aesthetics, financial structuring (lenders and equity partners), building onsite and offsite infrastructure, transportation and roads, complying with codes and regulations, marketing, public/private sector partnerships, cost and quality controls, and managerial execution.

The real estate development process is almost never linear; rather it is dynamic, complex, and ambiguous, requiring constant adjustment and rethinking, especially in locales where the permitting and regulatory processes can take years to unfold. Market conditions often change over the course of a large project's lifetime, meaning the developer has to have financial staying power and a commitment to the purpose and vision that drove the project's inception, but also an ability to adapt. Engineers and other professionals who are involved need that same understanding of the overall process and the changing conditions as well as the ability to adapt.

For decades if not centuries, there has been a tense debate in higher education between those who advocate for a liberal versus specialized education. Most of the engineers and technical consultants that developers work with today are products of the specialized school of thought. Successful developers and the engineers, architects and other professionals who work with them benefit from both technical expertise and an informed generalist's commanding view. Much constructive and creative work gets done before the final design and completed project ever emerges. Throughout the project those involved have to employ nimble, integrative thinking rather than blindly rushing to produce construction documents and drawings.

Tim Brown of IDEO (one of the most innovative design firms in the world) writes extensively about and is an advocate of a more evolutionary approach. IDEO employs a creative, open-ended design framework that allows the design of complex systems to unfold in a natural, almost organic manner. This results in better, more useful outcomes and, more often than not, true creative breakthroughs. Similar thinking is appropriate in real estate development.

Some colleges and universities are awakening to the need for engineers, lawyers, architects, doctors, and other highly specialized professions to gain a more complete yet basic understanding of market forces, business considerations, and economics. Business classes are being integrated into legal, engineering, architecture, and medical degree programs.

Dewberry's book is timely and in step with these educational reforms. He uses the concept of milestones to describe critical points in the development process continuum. At each of these milestones, he outlines the myriad of professional expertise needed and the roles each discipline plays. This book will help engineers, architects, and other technical professionals better understand their client's needs and the complex processes and decision points in the life of a real estate development project. Read this book and prepare yourself to better serve the needs of your developer clientele.

Gary W. Maler, MA
Director of the Real Estate Center
Mays Business School
Texas A&M University

PREFACE

When this business was launched in 1956 the Land Development Planning, Engineering, and Surveying professions were largely viewed as a backwater branch of Civil Engineering and not respected as a legitimate engineering field. Other consultants looked down their noses at anyone engaged in this practice and felt it was not "real engineering." Since land development consulting was how I made a living, I strongly resented the notion. I felt then, and still feel now, that this is a very noble profession. It requires expertise in all branches of civil engineering including surveying, roadway design, grading, drainage, water systems, wastewater systems, dry utilities, and environmental science; as well as knowledge of the related fields including urban planning, landscaping, archaeology, and architecture. We also have a responsibility to understand the economics, schedule, and vision of a project. But more important than the experience gained as a land development consultant, is the product of our diligent labor: we work to improve our communities.

For this reason, I have devoted my career to elevating this profession to the level it deserves.

In the early days, infrastructure was often considered an inconvenience that reduced the overall budget of the project. Few regulations required adequate drainage, utilities, and other infrastructure to provide good, reliable access to housing, employment, healthcare, public transportation, and retail developments. We often clashed with our clients over these issues as we advocated for sustainable and resilient infrastructure that exceeded the minimum standards. Gradually, the localities mandated better infrastructure and improved environmental performance through enhanced standards and regulations. These requirements are still progressing and evolving today as evidenced by the tremendous strides taken in the Green Building and environmental movements. Today, most developers and localities acknowledge the importance of infrastructure investments.

The profession of Land Development Consulting is now recognized and respected among the engineering disciplines. Every major A/E consulting firm has a land development practice. It is taught in many colleges and universities and in some cases, as its own specialty track within the civil engineering program. Young people are aware of and attracted to the profession. They enter this field inspired, bringing with them new ideas, the most recent technology, and a youthful perspective on the world that challenges us "old-timers" to keep pace with the speed of learning, rise above convention and truly innovate for the benefit of our clients and our communities. I feel that our firm, in its way, has greatly contributed and remains attuned to this dynamic industry with the *Land Development Handbook* series.

The first *Land Development Handbook* began as a dream of mine many years ago. In the mid-eighties I committed to developing a literary resource that could be shared with the

Pictured left to right: Sid Dewberry, Jim Nealon, Dick Davis when the firm was known as Dewberry, Nealon & Davis. Leaning on Nealon's era 1935 airplane.

civil engineering industry. Rather naively, I thought a book was something you sat down to do and finished within a few weeks' time. How surprised I was to learn that it would take years. The first edition of this text, in fact, took 7 years from start to finish. When looking for interest from publishers, I was pleased that many were anxious to publish and distribute the book. We then entered into what has become one of the most treasured and unique business relationships I have formed over the years with a premier technical publisher, McGraw-Hill Publishing. After the first edition, McGraw-Hill told us they would want us to update the handbook every few years (if the book was successful). By their measures this book is a best seller for the industry and continues to serve as a great resource. I'm proud to see the book in the offices of our clients, in the hands of students, and on the shelves of other design firms.

This business and our communities have changed dramatically since the first edition was published. To evolve with these changes, we regularly update the book to capture changes in policies and identify ever-changing design processes. With this fourth edition we have also expanded the scope of the book to better represent the expanded scope of land development. New technologies, tightened economics, and more complex projects require a broader range of knowledge. This fourth edition of the *Land Development Handbook* has evolved into three books to focus on (1) business, (2) design, and (3) construction. Together, these texts are the *Land Development Handbook* series:

1. *Development of the Built Environment* is all about the business and economics of public and private projects. This book is meant to improve understanding and communication between consultants and developers/ owners to ensure greater success of projects.

2. *Land Development Handbook* continues to focus on both the process and the technical design of civil engineering. We also emphasize the importance of public, private, and community relationships and involvement. We can't design in a silo—everything is connected to the community and the environment.

3. *Construction Practices for Land Development* describes the construction and operations of a project. Our industry is quickly trending toward design/build processes. Design is influenced by construction and operation considerations, and the design team should be actively involved in the construction.

I want to personally thank everyone who contributed to the fourth edition. Having been through this process several times before, I know the success of this exciting Dewberry endeavor is due to the dedication of each team member. A new edition of a book is no small task, and the development of two new books is a monumental effort. This latest edition truly represents a corporate-wide effort, as nearly all our 52 offices have contributed in both large and small ways. This diverse corporate presence has yielded valuable insight and fresh perspective from across the country.

Our lead contributors, Cody Pennetti, Chris Guyan, Kat Grimsley, and Claire White have demonstrated tremendous dedication and passion toward a shared vision of creating a great resource for the industry. This team is unique with backgrounds in both professional consulting and academia that adds depth to the content of the texts.

- Cody Pennetti served as an editor and contributor for this edition of the *Land Development Handbook,* and capably handled the complex task of managing the production of the three book *Land Development Handbook* series. Cody began his career with Dewberry and has always been passionate about the industry and continuous teaching and learning. Cody is now pursuing his PhD at the University of Virginia with a goal of serving in academia while staying involved in land development consulting. Cody lent a unique perspective and new ideas in this undertaking and helped produce a great resource for both the professional and academic industry.

- Christopher Guyan was instrumental in the development of this edition of the *Land Development Handbook* and operated as both an editor and contributor. Chris joined Dewberry after earning his undergraduate degree in civil engineering from Penn State. He is currently working on his MS in Urban and Regional Planning at Virginia Tech. He has shown tremendous promise with his design work and his efforts on the *Land Development Handbook*. Chris has a strong focus on the planning and design associated with land development. This focus has helped shape some of the underlying themes in this fourth edition to emphasize the importance of planned development and the community.

- Dr. C. Kat Grimsley was the primary contributor, writer, and editor for *Development of the Built Environment*. This new text benefits from Kat's extensive knowledge of commercial real estate. She is the director of the Masters of the Real Estate Development program at George Mason University, has served at the U.S. Department of State managing an international development portfolio, and has private sector experience working on financial modeling and commercial transactions. Kat completed her doctorate at the University of Cambridge in the United Kingdom with a focus on tenure security and international property rights and has since been appointed as a NAIOP Distinguished Fellow and member of the UN Economic Commission for Europe's Real Estate Market Advisory Group under the Committee for Housing and Land Management.

- Claire White was the primary contributor for the *Construction Practices for Land Development*. Claire earned her bachelor's and master's degrees from Virginia Tech and has experience working in the consulting field, starting as an intern with Dewberry. She has recently transitioned into a teaching role at Virginia Tech where she focuses on land development and real estate courses. From her consulting and academia roles, Claire has witnessed the critical role of engineers and the development team during the construction phase of a project. She has structured the text to help engineers improve design and further contribute to project success by anticipating construction processes.

Thank you all for your commitment to this endeavor! I also recognize that the efforts of these individuals are supported by family and friends that work behind the scenes. I want to sincerely thank those that are always there to support us while we work on these demanding projects. I have continued to hold the role of editor-in-chief for the fourth edition of the *Land Development Handbook* and the two new books in the series and have been proud to work with this team.

I also want to extend a special thank you to Dottie Spindle, my administrative assistant, who took care of the little things, the big things, and everything in-between so that I could focus on the things that truly matter to me, like these books and this company. Keeping me on schedule and on task is a challenge, but it is one she embraced with a smile for 33 years up until her recent retirement. Her replacement, Janice Spillan, a career executive assistant, has picked up the pieces seamlessly and continues to move the ball forward. Many thanks to her as well.

Peer reviewers are a critical component of our text. Those who think writing is difficult should try peer reviewing (or editing); balancing criticism with encouragement is a tall order. Our peer reviewers rose to this task under tight timeframes and across great distances. Their expertise was invaluable and that they were willing to lend it to this endeavor speaks highly of their commitment to Dewberry, to their practice, and to mentoring others.

I would like to thank Mark Hassinger and Matthew McClelland, who served as external reviewers for Development of the Built Environment. Mark is an industry expert and Partner at Hassinger/Armm Associates with more than 30 years of real estate development and management experience; he also serves as an Adjunct Professor of real estate development at George Mason University and is a founding partner of RE3—a real estate executive education provider, making him uniquely positioned to offer advice on the creation of a new development textbook. Matthew brought a business strategy perspective to the project, ensuring we incorporated best practices beyond just those commonly used in the development field; he holds an MBA from the University of Chicago Booth School of Business and is a Senior Manager of Strategy and M&A for Accenture. I am also grateful to the dedicated individuals recognized in the Contributors section, who each brought subject matter expertise to different chapters of the book. Their contributions have added depth that will benefit readers from all sectors of the development industry.

I would like to extend my sincere gratitude to all our clients for your continued support. Many of you have willingly offered components of your projects for inclusion in the text and we are happy to have your cooperation in this unique project. In particular, I would like to thank Milt Peterson, Founder of the Peterson Companies, for allowing us to include a detailed case study of his National Harbor project in *Development of the Built Environment*. National Harbor was, and continues to be, a monumental undertaking worthy of study by those in the development industry.

Two other individuals that have contributed greatly to this edition are Matt Pennetti and Dave Huh. Matt is an artist who developed hundreds of new illustrations and had the complex task of creating graphics of technically complex topics. Dave is a talented photographer at Dewberry who has an amazing eye for showcasing our projects in the best way possible. Our profession relies heavily on communication through graphics (plans, drawings, details), and these two professionals have contributed amazing works of art that complement the technical content of the books.

Craig Thomas, Dewberry's General Counsel, helped us to initiate this project with McGraw-Hill, has overseen all the contractual arrangements since we first published the *Land Development Handbook* many years ago, and has been a valuable legal resource throughout. Thank you, once again, for your support in this endeavor.

Finally, I want to express my deep regard for our partner in this effort, our publisher Lauren Poplawski of McGraw-Hill. She took over a large project and has been instrumental in the development of the new edition of the *Land Development Handbook* and the expansion of the *Land Development Handbook* series. Thank you for believing in us, for helping us elevate land development consulting as a profession and making one of my dreams—this book—come true, again!

In 1956, if you had told me that our six-person land development consulting company would grow into becoming one of the top 50 A&E companies in the United States, I would have thought you were nuts! I learned the hard way that real estate development was subject to the many ups and downs of the economy. For that reason, we sought early on to diversify our company into other facets of the A&E business. This diversification effort has been hugely successful for us, but land development continues to be one of the primary underpinnings of our practice. We love it and every new project continues to get the enthusiasm and professional care that we gave when we were first trying to get established. I would urge the thousands of small land development consultants throughout the United States to always consider the connection between the projects across geographies, design firms, the environment, and communities. It's necessary to diversify our skillset to adapt to industry changes and economics while also considering the big picture of our responsibilities as design professionals. And ultimately, deep down, we get supreme joy out of helping plan and build safe, healthy, financially feasible, sustainable, and beautiful places for people to learn, work, worship, shop, play, and live.

When I was born, 91 years ago, the horse and buggy had almost been completely replaced by the horseless carriage. Telephone, radios, and electricity were getting to be the norm. Television was just coming on the scene. The information technology era was being developed. Population in the United States had grown from 120 to over 328 million today. Air travel was beginning to be a serious competitor of rail, road, and water. The United States had been completely rebuilt. It is now a different place—totally new in just one lifetime. My first great grandchild, who is 18 months old, will see even more rapid changes with driverless cars and intelligent machines taking over our day-to-day activities, along with changes in healthcare. All these things will surely be a challenge to the engineers and scientists to provide solutions. Just in the next few years, many of these changes will have a huge impact on society, and how the engineers and scientists provide solutions for them will be amazing. Oh how I wish I were a young engineer facing these opportunities to provide solutions to the many challenges!

Sidney O. Dewberry, PE, LS
Chairman Emeritus, Dewberry
Editor-in-chief

CONTRIBUTORS

Editors and Primary Contributors:

Sidney O. Dewberry, PE, LS
Editor-in-Chief

C. Kat Grimsley, PhD cantab
Editor and primary contributor

Cody A. Pennetti, PE
Managing Editor

Additional Contributors:

Marty Almquist
Commercial Real Estate Broker
Meany & Oliver Companies, Inc.
Alliance

Bob Chase
President Emeritus
Northern Virginia Transportation

Karen Cohen, J.D.
Shareholder
Vanderpool, Frostick & Nishanian, P.C.

Melina Duggal, AICP
President
Duggal Real Estate Advisors, LLC

Greg Hoffman
Director of Development
Rooney Properties

Roger Lin, J.D.
Managing Partner
Eyremount Investments

Lauren McCarthy
Program Manager
Center for Transportation Public-Private
Partnership Policy
George Mason University

Benjamin Myers
President
Myers LLC

Ronaldo T. Nicholson, P.F.
Vice President
Parsons

Huong Van
Senior Vice President
FVCbank

Malcolm Van de Reit II
City President, Mid-Atlantic region
LMC, Lennar

Ian Whitehead, P.E.
Senior Project Engineer
ECS

Reviewers:

Mark Hassinger
Partner
Hassinger/Armm Associates

Matthew McClelland
Senior Manager, Strategy and M&A
Accenture

Dewberry Contributors:

Tom Christensen, PE
Associate Vice President

Timothy C. Culleiton, PE
Associate Vice President; Civil Engineer

Bill Fissel, PE
Senior Vice President; Civil Engineer

Christopher J. Guyan
Civil Engineer

Aileen Heberer
Graphic Design Manager

Dave Huh
Photographer

Molly Johnson
Director of Communications

J. Paul Lewis, AIA
Senior Architect

David J. Mahoney, PE
Executive Vice President; Transportation Engineer

Kimberly McVicker
Communications Manager

Beth Patrizzi
Staff Environmental Scientist

Matt A. Pennetti
Illustrator

Craig N. Thomas
General Counsel Dewberry

Kurt Thompson, PE
Executive Vice President; Infrastructure

J.I. "Jack" Vega, PE
Vice President; Civil Engineer

Claire M. White, PE, ENV SP
Assistant Professor of Practice, Virginia Tech

DEVELOPMENT OF THE BUILT ENVIRONMENT

CHAPTER 1

DEVELOPERS AND THE DEVELOPMENT PROCESS

As real estate development has evolved to accommodate increasingly more complex regulation and sophisticated built structures, so too has the role of civil engineering evolved. Historically, civil engineers focused chiefly on the technical components of basic design and construction; however, they and other development team members now participate actively in consulting throughout the entire lifecycle of modern development projects. This expanded role often means that contemporary civil engineers may find themselves facing contentious, cross-disciplinary decisions that their predecessors did not have to address. This textbook will explore the entire development process from an applied perspective, rather than a technical one, to provide a better understanding of the business considerations that influence development projects at different stages. This will contribute to a holistic understanding of the development process and will allow civil engineers and other development team members to better assist their developer-clients at key stages throughout the project.

1.1. DEVELOPMENT AND DEVELOPERS

Real estate development is a multifaceted, inter-disciplinary endeavor capable of producing some of the highest returns of any investment activity. However, though it can be extremely profitable, real estate development is both complex and costly. Even simple development projects can be multi-year undertakings fraught with risk, unexpected challenges, and setbacks. For purposes of this book, *development* is defined as the process of finding a site, determining what should be built on it, obtaining funding for the project, getting permission to build, overseeing construction through to completion, and recognizing financial return either through sale of the asset or ongoing operations. Development often involves transforming virgin land into something else, usually infrastructure or a new building. However, the term can also include *redevelopment*, which involves the extensive remodeling, demolition, or replacement of an existing building. It can also refer to *infill development*, which pertains to building on underutilized or vacant urban lots that are surrounded by other buildings or structures.

The person or entity that directs a specific development project is often referred to as a *developer* or *development director*. Private sector developers may also be referred to as *sponsors* by their investment partners, as *borrowers* by lenders, as *clients* by members of the development team, and as *landlords* by their future tenants. Note that for financial and liability reasons, technically most developers are actually corporate entities. However, an individual is always responsible for leading each development project and guiding the development team. In a small firm, this may be the principal, who has considerable authority over decisions; in a larger firm, this may be the firm's designee or project manager, who may have to report certain decisions to an investment committee for approval. For the sake of simplicity, this book will refer to the "developer" as the individual person, whether principal or designee, guiding a development project.

Developers are supported by an extensive interdisciplinary *development team* comprised of (but not limited to) technical experts such as civil engineers, environmental scientists, architects, and attorneys. For any development project to be successful, it is important for all development team members, including the developer, to understand not only their own roles but also how the roles of the entire team are interdependent. The degree to which the team successfully communicates and balances their roles is directly related to the success of a development project.

This chapter will begin by discussing the typical characteristics of developers, examining their role in the development process, and outlining different kinds of development models. It will then provide detail about core members of the development team and highlight areas where team members may inadvertently find themselves in conflict with each other or their developer-clients. Finally, this chapter will provide an overview of critical milestones in the development process.

1.1.1. Public Sector versus Private Sector Developers

Developers will pursue different types of projects depending on their individual preferences, risk tolerance, return requirements, and available resources. Perhaps the most important distinction to make is between developers operating in the private sector versus the public sector. The *public sector* refers to the government and its agencies and departments. While private sector developers are individuals or corporations, public sector developers are government entities at either the federal, state, or local level. Although most people tend to think only of private sector developers, public sector development represents a major component of

the development industry. Examples of public sector development projects include large-scale infrastructure projects, such as highways, as well as building projects such as post offices, courthouses, government-owned agency office buildings, and local schools or state universities.

The fundamental development tasks are essentially the same for both public and private sector developers: they must find and evaluate a potential site, obtain funding, and coordinate the design and construction of each project. To do so they must both interview and hire technical experts who form their development teams. Additionally, both public and private sector developers must also be knowledgeable enough about all aspects of the development process to play an active role in coordinating their teams' activities. Finally, public and private developers are each responsible for managing the expectations of their stakeholders both before and during a project, especially as unforeseen circumstances arise. Despite these high-level similarities, however, the priorities and processes of public and private sector developers are extremely different.

Private developers seek to achieve financial returns from their investment in projects, whereas the goal of public sector developers is to provide services to citizens. Thus, differences in the public process are evident from project inception, where new development is *not* triggered by discovering an opportunity nor driven by the desire to earn profit. Rather, it is often a somewhat sterile function of evaluating public sector needs and current capacities, appropriating necessary funding, and gaining permission to begin a formal procurement process. This is not to say that the public sector necessarily lacks innovation in its projects or that individual government employees are not passionate about their work. It is simply that the mechanisms guiding public sector development are generally inflexible. Perhaps unsurprisingly, as a developer the public sector is risk averse and rule oriented. Public sector developers are accountable to tax payers and political leaders for their results and, as such, are required to comply with a multitude of laws, regulations, and policies that guide everything from procurement to construction practices. The federal, state, and local regulatory structure is complex and prescriptive, including components that are sometimes overlapping or interdependent.

These controls are intended to ensure responsible stewardship of tax revenue, fair competition, and transparency in government operations; however, somewhat ironically, this cumbersome structure often makes the public sector process less efficient and more costly than that of the private sector. The unique challenges of public sector development will be discussed in Chap. 7, although each chapter will point out key differences between private and public sector development projects as appropriate.

In contrast, private sector developers are incentivized by profit-seeking activities, which can make them appear to be unsupervised opportunists when compared to their public sector counterparts. Unlike many development team members, developers are not licensed and are constrained only by their own partnership relationships or sometimes, for larger development firms, investment committee approvals. They tend to be optimistic and can have a fairly high tolerance for risk, calculated or otherwise, as compared to other professions. Many, although certainly not all, private sector developers are good negotiators and tend to have confident, outgoing personalities. These are all important leadership characteristics for the private sector developer, who must usually put his/her own time and money into a deal and is only rewarded if the project is successful. Investors and industry professionals are unlikely to work with a developer unless they respect his/her integrity, so establishing a reliable and trustworthy reputation is extremely important for developers. Networking skills are understandably important, and developers often participate actively in industry organizations that offer both informative events and networking opportunities, such as:

- The Commercial Real Estate Development Association, which is known by the acronym "NAIOP"
- Urban Land Institute (ULI)
- International Council of Shopping Centers (ICSC)
- Building Industry Associations
- Commercial Real Estate Women (CREW)

At the most basic level, much of what private sector developers actually do is seek opportunity and solve problems. Fundamentally, an initial opportunity comes in one of two forms for developers: (1) finding potential sites to develop, or (2) finding clients who will become future tenants or buyers. In other words, as captured by a well-known industry saying, in order for a new project to begin, a developer must find either "a site looking for a use or a use looking for a site." It is the developer's responsibility to identify such an opportunity and then leverage his/her capabilities and contacts to bring the necessary parties together and create a development project.

Unlike public sector developers, who are government entities with no mandate to earn profits, private sector developers typically earn a living in four ways: (1) by charging fees for their work, (2) by creating value and selling properties to investors after they are built or entitled, (3) by receiving rental payments from tenants occupying properties they build and continue to own, and (4) by capturing the appreciation in the value of their ownership position in properties they build and continue to own. In virtually all cases, profit cannot be realized until after a development project is completed, meaning that during the development process itself, developers must take the risk that unexpected factors will reduce or eliminate their future earnings. In order to safeguard future returns, developers will constantly seek to control project costs and timing. Anything that increases either of these two critical factors will weaken the project's overall profitability and, in serious circumstances, can affect the developer's ability to repay investors and lenders. The sooner a project is completed, the sooner the developer can begin to generate profit.

1.1.2. Types of Private Sector Developers

Developers will pursue different types of projects depending on their individual preferences, risk tolerance, return requirements, and available resources. This section will describe common types of developers. Note that these descriptions pertain to private sector developers; public sector developers will be explored more thoroughly in Chap. 7, which is dedicated to the importance of public sector development.

Merchant Builders. A developer who builds with the intention of selling the asset immediately upon (or even before) completion is a *merchant builder*.

Essentially, all homebuilders are merchant builders because they intend to sell the houses they build as quickly as possible. However, merchant builders can develop any type of property, including office, retail, or industrial. The merchant builder's profit is maximized by completing development projects as quickly and cost-effectively as possible. In the commercial realm, this means that a merchant builder may make different choices about building systems or finishes than other types of developers who intend to own and operate a property over a long term.

In some cases, a merchant builder will have identified a buyer before starting construction. This is often the case when a firm wants to own a new building but does not have the competency to undertake the development itself. In this instance, the firm contracts with the developer to design and construct a building to their specifications in exchange for a future purchase. This is known as a *build-to-suit* project.

Not all merchant builders have buyers in advance of construction. However, they may have identified a large tenant willing to rent space in their proposed project. Having such a tenant sign an advance lease for space in the future building is known as *pre-leasing*. The lease, which represents a future income stream, allows the merchant builder to seek financing for the project; without such a guaranteed source of revenue, lending institutions will often not take the market risk associated with financing a potentially vacant building. Depending on the strength of the merchant builder, as much as 50 percent pre-leasing is necessary to obtain funding and move forward with construction. During construction, the merchant builder will continue to seek additional tenants as well as an investor to purchase the completed building.

There are several risks and rewards associated with this business model. The merchant builder realizes a fairly quick return by selling as soon as possible and is able to immediately transfer his/her profit into another project. This model is also efficient in that it does not require the developer to engage with the liability and management-heavy burden of owning and operating a property. However, merchant builders also face risks and are susceptible to short-term changes. For example, if market conditions change, it may not be possible to find either tenants or a buyer. In a worst-case scenario, tenants with existing pre-leases could default and not take occupancy, leaving the merchant builder as the reluctant owner of a partially (or completely) empty building generating little in the way of rental revenue and net profit. Even when the merchant builder is able to successfully pre-lease and build, he/she bears the risk of being able to find an investor willing to acquire the asset at the required minimum asking price.

Developer/Owner. A developer/owner is one who intends to retain ownership of a property after construction is finished and tenants have taken occupancy (at which point the developer becomes a landlord). Public sector developers generally fall into this category, although their motives are different than those of their private sector counterparts. The motivation for most private sector developer/owners is to create a passive income stream in the form of rent that they can collect for many years in the future. They also capture appreciation in asset value over time. Most developer/owners estimate a 10-year holding period for purposes of their financial analysis; however, there is no "correct" hold period and not all developer/owners use the same investment horizon. Certain developer/owners, especially family-owned firms, focus on longevity and may hold a well-located asset for 30 years or more, giving future generations the opportunity to redevelop or sell it. Note that the developer/owner's intent to retain ownership of a property certainly does not preclude him/her from selling the asset if it makes sense to do so.

In general, developer/owners have a vested interest in creating efficient and resilient buildings that will be less costly to operate and maintain. This means that they have an incentive to install higher-quality, but usually more expensive, systems or fixtures that will function better over time. This can increase up-front project costs in favor of reducing the cost of future operations, which must be accounted for in the developer's financial projections.

Like merchant builders, developer/owners often engage in build-to-suit projects. However, in these instances the developer/owner does not sell the finished building but, instead, becomes the landlord while the build-to-suit client becomes the tenant. Developer/owners can also build speculatively, sometimes referred to as building "spec," with the hope of leasing out their buildings during the construction process or shortly after completion. A developer's ability to do this will depend on market conditions and his/her lender's pre-leasing requirements.

Developer/owners face several risks, although these are often different than those faced by merchant builders. For example, developer/owners must invest a large portion of capital in a single property for many years, limiting the cash available to use for the next project. This creates liquidity risk and opportunity costs for the developer should future opportunities be more lucrative than his/her current investment. Although all developers look to offset their required contribution with external funding sources, most developers are ultimately required to make at least some direct investment themselves. Developer/owners must also consider market risk. As owners, they are subject to market conditions beyond their control that may affect future tenant-demand for their building(s). Finally, developer/owners bear the liability and responsibility of managing an operating property. Some developer/owners hire a third-party management firm to serve this function while others have their own internal management division. Hiring a third-party manager is more costly, but self-management is more time consuming.

Institutional Developers.
In the context of commercial real estate, the term "institutional" is normally associated with very large corporate investors. However, the term can also be applied to very large, publicly traded institutional development firms. These tend to be real estate investment trusts (REITs) with national or international portfolios. Their level of sophistication, project scope, and overall capabilities generally surpass those of smaller, local developers/owners. However, they often share the same business model of developing and retaining ownership of their properties.

Small-Scale/Entrepreneurial Developers.
On the opposite end of the spectrum from institutional firms are small, entrepreneurial developers. These may be single individuals or small groups of partners. They may focus on a merchant building model or develop with the intention to retain ownership. The biggest challenge facing entrepreneurial developers comes from their lack of resources. Unlike their corporate counterparts, entrepreneurs do not have in-house staff to help them complete the various tasks required for

development, such as market analysis, financial evaluations, construction pricing, or legal review of contracts. Entrepreneurs usually pursue smaller deals and may complete fewer overall projects due to their resource constraints. However, perhaps as a result of need, entrepreneurs can also be especially innovative in structuring deals and investment arrangements in order to make projects possible.

Of course, innovative and entrepreneurial behavior is not exclusively the domain of smaller firms. Large firms can also take entrepreneurial approaches to design, project delivery, and other elements of development. However, for purposes of this book, a developer who is an "entrepreneur" is assumed to be the owner of a smaller development firm.

Specialty Developers.
As with many professional fields, developers can become specialists in a particular area. This tends to be driven by a developer's preferences and experience over time. Specialty developers usually focus in one of two areas: either (1) on a particular type of building, or (2) on a certain segment of the development process.

Developers that concentrate exclusively on certain types of buildings are said to specialize by product type or asset class. *Asset class* refers to the relative quality of a building based on age, amenities, and other factors. *Product type* refers to the categorization of buildings based on their primary use, with the main product types being: residential, retail, office, and industrial. Residential can be further subcategorized as either single-family or multi-family, depending on the nature and density of the projects. Developers who specialize by product type often have the advantage of a well-formed niche competency including specialized banking and equity relationships and a thorough understanding of the competitive environment for their particular product type of choice. However, they also face certain disadvantages such as lack of portfolio diversity and lack of flexibility in weak markets.

Not all firms involved with development have an interest in completing the entire process. Some specialty developers focus exclusively on one portion of the process and develop niche competencies in that specific area. They then use their abilities to add value at their stage of the process before reselling a property to another developer. For example, a specialist developer might purchase several small parcels of land and retitle them into a single, larger piece of property. This is known as *assemblage*, but may also be referred to as a *consolidation*. The developer can then sell the new, larger parcel at a premium because it can accommodate a larger development project and saves a future developer from having to spend time and money creating the assemblage. Conversely, a specialist developer may take a large piece of land and go through the formal process of dividing it into individual lots. This process is called *subdivision* and the legal mechanism by which individual lots are created is called *platting*. After subdivision, the larger property can also be sold at a premium, for example, to a developer specializing in homebuilding who is willing to pay more for a site that is already subdivided and ready for home construction. Other developers

specialize in rezoning and entitlements, which are complicated, risky, and lengthy processes.

Different product types and other forms of specialized development will be discussed in Chap. 3. The rezoning and legislative approval process will be addressed more fully in Chap. 6.

1.2. THE DEVELOPMENT TEAM

This section will describe the roles of key members of the development team and highlight areas where team members might encounter friction with their developer-clients or each other. Professionals from many different disciplines are involved in any development project, which often leads to the practice of classifying members of the development team into different subject matter teams as represented in Fig. 1.1. Note, however, that these different professionals do not necessarily work

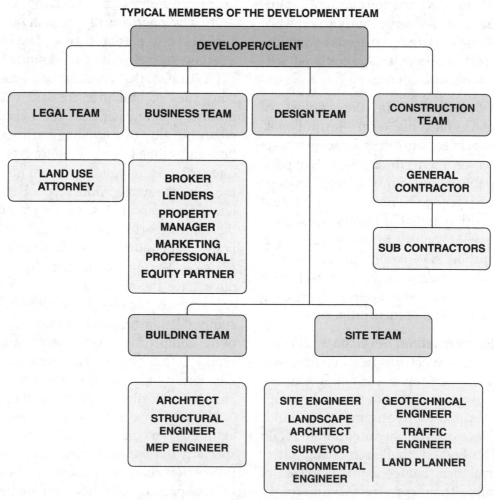

FIGURE 1.1   Typical members of the development team.

together under the same firm or even the same service contract. Depending on the type of project delivery method used, the developer may contract with a unified team under a single contract or with different firms independently for different services, even within a particular "team." Prioritizing clear communication both within and across teams is critically important, especially under disaggregated contracts or delivery methods.

1.2.1. Design Team

The *design team* is comprised of engineering and architecture experts that are responsible for creating site and building designs. These professionals can be further divided into a building team and a site team. As the names suggest, the *building team* is responsible for vertical structures while the *site team* focuses predominately on horizontal site work. Both teams are involved in situating a proposed project on a subject site and contributing to the project approvals process under the guidance of the developer with input from market and finance subject matter experts.

Engineers.   Engineering is a broad professional field of technical experts. The types of engineers most commonly involved in development projects are civil engineers. *Civil engineers* are technical professionals who "design, build, supervise, operate, and maintain construction projects and systems in the public and private sector."[1] They also provide their developer-clients with estimates of the costs for labor and materials of the project. Civil engineers can specialize in one of several areas related to development, including site development, geotechnical, transportation, structural, water resources, or environmental. Each of these specialized areas is critical to the success of a development project. It is worth noting that both public sector and private sector developers require the same professional services from engineers and largely differ only in their bidding and contract requirements.

Civil engineers are not required to be licensed for early career level jobs; however, they may wish to obtain a Professional Engineering (PE) license in order to move into management roles or work on large, complex projects.[1] The PE license is issued at the state level, but generally requires an engineer to hold a degree accredited by the Accreditation Board for Engineering and Technology (ABET), pass two separate exams issued by the National Council of Examiners for Engineering and Surveying (NCEES), and have four years of work experience. When taking the PE exam, engineers will pick an area of specialization within the field of civil engineering and will have to pass a specialized portion of the exam.[2]

Though it is not a required license, many engineers choose to become certified in Leadership in Energy and Environmental Design (LEED). LEED certification allows them to assist their developer-clients in meeting the U.S. Green Building Council's (USGBC) standards. Note that the USGBC is *not* a government entity. Most civil engineers are also members of the American Society of Civil Engineers (ASCE), which is the primary professional industry organization for civil engineers.

As development team members, civil engineers are valued for their advanced technical and mathematical skills as well as their high aptitude for quantitative problem solving, particularly with site challenges. However, they can come into conflict with their developer-clients precisely because they are so technically oriented, whereas the developer is focusing on cost, timing, and overall project feasibility. Civil engineers are often not fully aware of the financial, schedule, and other pressures facing the developer. Similarly, developers are often not fully appreciative of the technical engineering requirements that may pose constraints for their civil engineering team members.

Example: Architect-Engineering Conflicts

In many projects, there is an overlap between the design responsibilities of the building team and the site team. While the entire design team works together to create the finished product, it is important that each discipline be aware of the limits of their own responsibilities and the need to accommodate the priorities of other team members' disciplines without undue resistance, which can be counterproductive. For example, the team architect may present site layouts at early design stages without being fully aware of all site requirements that modify the form and function of the site. While the architect may set the initial building location on a site based on aesthetic considerations, the civil engineer may subsequently identify the need for adjustments based on zoning requirements or geotechnical conditions that disrupt the architect's vision for practical/cost-based reasons. Similarly, most architectural renderings and models show a "flat" site, but actual topographic conditions, which can impact the aesthetics and visibility of the building, must be considered from the engineering perspective, especially where compliance with the Americans with Disabilities Act (ADA) may be affected. Further, the building team and site team likely have different schedules for their respective permit processes, and when each discipline is focused on their own work to the exclusion of others, errors and omissions can easily occur that affect the greater effort, including:

- Utility connections between building and site systems may have discrepancies in location, size, and material.

- Door locations and accessibility requirements may be omitted.

- Exterior grades along the building face may not work with building facade or windows.

- Height measurements are often different in building codes than in zoning codes, which can cause confusion for building requirements.

Despite challenges, design team interactions can be successfully managed through deliberate communication and periodic update meetings. Often, innovative project solutions can be achieved when team members collaborate effectively.

Architects.   Architects design spaces for use and occupation. Their projects can range from small interior office layouts to the rehabilitation of an existing building to the complete concept creation and design of a new building. Architects are involved in several different stages throughout the development process, supporting the developer as

he/she looks for suitable sites, refines the project scope, and ultimately settles on a final design. The developer's final construction documents must be approved by a licensed architect. As with engineers, both public sector and private sector developers require professional architectural services but, again, differ in their bidding and contract requirements.

Architects must be licensed to practice. Architecture licenses are issued by each state according to individual state standards; however, virtually all states require, at the minimum, an architect to demonstrate competence via three criteria: education, experience, and examination.[3] The education criteria can be satisfied with a bachelor's or master's degree in architecture from a university accredited by the National Architectural Accrediting Board (NAAB). An aspiring architect can obtain the necessary experience through the National Council of Architectural Registration Boards (NCARB) Architectural Experience Program™ (AXP™), which requires he/she log more than 3,700 hours of work, at least half of which must be under the direct supervision of a licensed and practicing architect. Finally, an architect must pass a national exam, the Architect Registration Examination® (ARE®). Licensed architects might also have an NCARB Certificate.

If there are additional requirements set by the particular state in which an architect wishes to practice, then he/she must also meet those requirements in order to become licensed. There are also rules about whether or not a licensed architect is legally permitted to practice in states other than the state where he/she was originally licensed, unless an additional license is obtained in that state.

The two main industry organizations for the architecture profession are:

- The American Institute of Architects (AIA)

- The National Council of Architectural Registration Boards (NCARB)

For development projects, architects can either be hired directly by a developer or as part of a team that includes engineers. If engineers are included in the team, they will be contracted through the architecture firm. The fee structure for architecture services is dependent on the type of project being completed and is specified by contract. Architects typically to use one of the many standard architectural services contracts provided by the AIA.[4] Different contract forms are used under different circumstances depending on the size and duration of the project and whether or not the architect is able to make a reasonable estimation of the cost to complete the required work. For example, if the architect has negotiated a fixed fee in exchange for a particular set of services, also known as a *scope of work* (SOW), for a large project, he/she would likely use the A101 "Standard Form of Agreement Between Owner and Contractor where the basis of payment is a Stipulated Sum." However, if the project is ambiguous or not yet fully defined, the scope of the architect's work might be unknown, making it difficult or impossible to estimate a fair fee. In this instance, the architect might use the A103 "Standard Form of Agreement Between Owner and Contractor where the basis of payment is the Cost of the Work Plus a Fee without a Guaranteed Maximum Price."

The architecture profession is an inherently aspirational and artistic one. Developers entrust their architects with the important task of *creating*, or taking the developer's vision for a parcel of land and helping render it into a viable building. Unfortunately, architects sometimes find that their artistic ideals are not supported by the economics of a deal. This can lead to friction wherein an architect wants to include certain design elements that would increase project costs, yet the developer may need to limit costs in order to provide investors with their required return. Conversely, a developer may falsely believe that an architect is simply "being artistic" and fail to appreciate important design recommendations that can materially improve the project.

Example: Design Team–Developer Conflict

Development design usually requires incremental improvement before the best layout is identified. During early phases of the design, the developer should work closely with the design team to accurately identify program requirements and site constraints as soon as

possible. Major design changes can impact the cost and schedule for a project if new requirements are introduced late in the design process. For example, late design changes can require substantial rework and will cause the developer to incur additional fees. It may also be expensive to accommodate changes to project components within the existing project design. Further, late changes combined with impending project deadlines increase likelihood that design errors may occur. However, site design does not exist for its own sake and must accommodate programmatic changes as they occur. In some instances, conditions beyond the developer's control may change and impact the project, leading to the need for design adjustments. This must be understood by the design team as part of the uncertainty inherent of development projects.

1.2.2. Business Team

For purposes of this book, several different professions have been included as part of the business team. However, there is less functional overlap between these professions than exists within members of the design team. For example, while architects and civil engineers often partner under a shared contract, lenders and commercial brokers do not share professional operations. This is not to say that the different members of the business team fail to work collaboratively; rather, they may simply not share contact, except through the developer.

Lenders. *Lenders* are entities that provide a source of cash in the form of a loan, also known as *debt financing*, for real estate investment and development projects. There are several different kinds of lenders. Two common examples are (1) commercial banks that finance loans using their own financial assets, and (2) institutional investors, such as pension funds or insurance companies, which invest their stakeholders' funds. All lenders, regardless of type, earn revenue by collecting interest on the amount of the money they loan. Just as there are different types of lenders, there are several different types of loans they can provide. These will be described in Chap. 5. Traditionally, developers seek debt financing from banks; therefore, this book will focus primarily on that sector of the lending industry and use the terms "lenders," "banks" or "bankers," and "banking lenders" interchangeably unless otherwise specified. Note that traditional lenders are not used by public sector developers, who primarily fund projects through appropriations from tax revenue or the issuance of bonds. In instances where members of the public sector and private sector form a development partnership, the private sector developer may seek loan financing.

Banks are licensed through a review process tied to either their state or national charter. Their operating activities are overseen by the Office of the Comptroller of the Currency (OCC) or Federal Deposit Insurance Corporation (FDIC). Note that individual bank employees do not require a license to practice. The major industry organizations of the lending profession in the banking sector include:

- American Bankers Association & State Associations (i.e., Virginia Bankers Association)

- Mortgage Bankers Association of America

- Federal Deposit Insurance Corporation (FDIC)

- Basel Committee on Banking Supervision

To receive a bank loan, the developer must submit a financing package to the bank containing specific details about the site, the project costs, and the anticipated financial profits. Additionally, the developer must also submit proof that he/she has the financial capacity to undertake the project. The developer's loan package will be examined by the lender through a process called *underwriting*, during which the lender will evaluate the project's overall risk and viability. If the lender believes that the project can be underwritten (or approved), the bank will proceed to negotiate loan terms with the developer.

Lenders operate in a binary environment: a loan is either approved or is not approved; a bank never partially lends money. By default, the lending profession is focused on forcing complex projects with many possible, nuanced future outcomes into a format that yields a strictly Yes/No lending decision. Sophisticated lenders may be able to create some room for creative terms in their approval process, but most lending requirements are fairly rigid and risk averse. This contrasts starkly with the developer's tolerance for risk and tendency to embrace possibility in order to bring a project to fruition. Therefore, lenders can easily find themselves in conflict with their developer-clients or other members of the development team.

Example: **Lender-Design Team Conflicts**

Two imperatives of commercial real estate lending are that loans close on time and have a fixed period of repayment. However, development is not a predicable field and meeting the bank's timing requirements can be problematic if a project is delayed or leasing has not occurred as quickly as predicted. Lenders often require that final engineering/architectural drawings and permits be included as part of a loan approval package, but these can be delayed because of design changes related to jurisdictional approvals, market conditions, design team conflicts, or other unforeseen circumstances. While the design team must make every effort to accommodate lender deadlines, this is not always possible.

Equity Partners. *Equity partners*, or *equity investors*, are private investors that lend money to developers in exchange for becoming partners in a development project and receiving preferred returns, which are higher than returns paid on standard debt financing. Equity investors can command this higher return because, unlike banks, these investors do not necessarily have any claims on the property title and, thus, take a higher risk of not recapturing their investment if the project fails.

Equity funds can be solicited at different phases of a development project, although they are most often sought once the developer is close to securing conventional debt financing or has a lending commitment. At that point, the developer has certainty with respect to how much equity is needed to bring the project to fruition. For example, if a bank will finance 75 percent of the cost of a project, the developer will bring together one or more equity investors to provide some or all of the remaining 25 percent needed to fund the project. Any shortfall in funding will have to be covered by the developer directly. Established developers may be able to raise a "blind pool" of equity, which refers to funds that are not tied to a particular project. This enables the developer to apply the money across different projects at various phases and with different risks and returns. Equity can also be assembled on a project-by-project basis, in which case the required rate of return will be based on the particular project. For example, equity funds applied to the early phases of a project for the entitlement process will typically require a much higher return because of the level of risk involved in entitlements. More information about equity contributions will be provided in Chap. 5.

Commercial Brokers. Commercial real estate brokers are licensed at the state level through the same process as residential brokers; however, that majority of commercial brokers work exclusively on commercial transactions, which tend to be larger and more sophisticated than residential home sales. Commercial brokers often specialize in either leasing space or selling commercial assets. They may further specialize in a specific type of commercial property, such as retail centers or office buildings. Brokers earn commissions based on the value of successfully completed transactions, including both leases and sales. In the case of development, this translates to the value of the land at acquisition or upon the negotiated sales price if/when the developer sells the future asset. Brokers are market experts that help their developer-clients by identifying sites and providing information on market conditions. A good broker will also be able to help the developer anticipate future (or forecasted) market conditions and discuss other planned development projects in the market. During a site acquisition, brokers can help negotiate the business terms of a transaction between their developer-client and the seller of

the land, so that the parties are agreed in principle before incurring attorney fees for contract negotiations. Once a development project is underway, a broker will help the developer identify tenants to lease the new space being built.

1.2.3. Legal Team

The legal team may include a single attorney or may include several different attorneys, depending on the scope of the project and the degree to which unusual legal issues arise. Large developers may use their own internal counsel but may also contract for specialized external legal assistance if necessary. Smaller entrepreneurial developers may be tempted to limit legal fees in order to conserve scarce resources but should always engage an attorney when necessary.

Attorneys. Attorneys specializing in commercial real estate have a critical role in the development process. Owners, developers, contractors, local governments, and other parties involved in real estate development retain lawyers to protect their legal interests and to help manage risk. When acting on behalf of a developer-client, an attorney may negotiate and draft contracts, conduct title examinations, provide zoning analyses, advise clients regarding the structure of a contemplated transaction, or appear before local governments to advocate for approval of a plan of development. This is not an exhaustive list of the many functions that may be performed by a commercial real estate lawyer. Additionally, when disputes arise and a party files a lawsuit, attorneys are hired to litigate and, where possible, the attorney can help the parties resolve such disputes through settlement (i.e., without the necessity of a court trial). Where an ongoing development project is in jeopardy, attorneys may assist the parties by drafting the appropriate documents for a "workout" arrangement between the various stakeholders.

Attorneys are licensed to practice law at the state level. Generally, the state's Supreme Court will empower an administrative body to determine licensing requirements and evaluate attorneys. A portion of the licensing requirement is always passing the state's bar exam. As part of their licensing, attorneys must be members of their state-level bar associations. They may also join voluntary bar associations, such as the American Bar Association (ABA), which is the primary national-level industry organization for the legal profession. Note that the ABA is not involved with the licensing of attorneys, although it is the accrediting body for university law programs issuing J.D. degrees.[5]

Unlike engineers and architects, whose services may be paired in large design-build or design-bid-build scenarios, attorneys tend to work in an individual capacity with their clients. Typically, their services are billed hourly; however, some also charge a graduated "flat fee" with a set fee for certain phases, or a hybrid of flat and hourly (e.g., a set fee for producing the first draft of a contract, then hourly thereafter) with invoices sent directly to the client. At the beginning of a new project or new relationship with a developer, attorneys may also require the payment of a *retainer*, which is a set amount of money to be paid in advance by a client.

Example: Attorney-Developer Conflict

Attorneys are trained rigorously to be detail-oriented professionals. This focused approach is essential to help developers navigate complex real estate transactions where there are usually multiple contracts, many of which can be quite complex. However, hiring an attorney does not relieve the developer and other represented parties of the duty to read and understand the agreements they are signing. Lawyers use their knowledge of the law to explain legal risks to their clients and to advise how to mitigate those risks. The client, however, makes the business decisions and determines what level of risk is acceptable. Depending on the nature of the legal issue involved and the level of risk associated with it, there can be a tension between the developer-client's desire to "get the deal done" and the lawyer's efforts to minimize risk to a level that the client indicates is acceptable. For this reason, developers generally value lawyers who are nimble and creative problem solvers. Rather than being a roadblock to closing a deal, lawyers are often most effective when they offer their developer-clients

options for how to achieve the ultimate business objective, while complying with the law and minimizing risk.

1.2.4. Additional Team Members

Although not covered in detail in this chapter, a host of other team members are integral to the development process. For example, virtually all projects require title reports, appraisal reports, traffic studies, geotechnical studies, and environmental studies, which are provided by specialty consultants. Larger, more complex projects often encounter unique challenges and can require even greater specialist support. The following is a non-exhaustive list of additional team members. The contributions of many of these specialists will be discussed in future chapters throughout this book.

- Builder/general contractors
- Traffic consultants
- Geotechnical and soils specialist engineering firms
- Environmental specialists
- Wetlands specialists
- Hazardous substance consultants
- Appraisers
- Place-making consultants
- Life safety and fire protection consultants
- Acoustical engineers
- Specialized lighting consultants
- Urban planners
- Landscape architects
- Property managers
- Marketing professionals
- Graphic designers
- Title abstractors

1.2.5. Team Communication and Conflict Management

Solving problems, balancing trade-offs, and resolving conflicts are daily occupations for developers as they strive to shepherd each development project from inception to completion. While most challenges are site or approval related, developers do sometimes come into conflict with members of their own development team. The causes of these internal conflicts are often tied to communication issues.

With respect to personality, the same traits that make private sector developers good at their jobs can also be liabilities when interacting with other team members. For example, developers tend to have strong, confident personalities, which allows them to generate support and lead a project. However, their macro focus on "the bigger picture" means that they may downplay the seriousness of important—and very real—problems or assume all issues can be solved if the development team simply tries harder. This behavior tends to be especially prevalent in cases where the developer believes a particular kind of problem was solved on another project, even though, in reality, the specific technical circumstances may not be comparable. By not heeding advice from the development team, developers can be overly confident or too willing to embrace unnecessary risk, which can have disastrous effects on a project. Similarly, developers can also be at risk for believing they understand more about issues than their technical experts. This can cause a developer to react poorly when he/she does not like solutions presented by the development team. However, rather than rejecting work or demanding it be redone, it is important for a developer to trust the experts he/she hired and seek clarification for why certain recommendations are being made. It may be that the proposed solution, while not to the developer's liking, is being offered for good reasons. For example, it could be more efficient or cost-effective in ways the developer has not considered. Equally, if the developer has not fully communicated the *reasons* behind certain requests, the team has little hope of successfully accommodating the developer's unknown underlying needs. If inexperienced developers push too hard for the impossible, they can alienate reputable professionals in other sectors of the industry and damage their own reputations as reasonable professionals.

Nonetheless, it is important for members of the development team to recognize that the developer is their client and actively seek clarification when necessary. Further, technical experts must recognize solutions or processes that appear desirable from their individual perspectives may not actually be helpful in the larger context of a development project. Good team members must trust their developer-clients, just as good developers must trust their teams. All parties should prioritize clear and continual communication about potential issues as soon as they arise.

The following are examples of areas in which well-meaning developers and team members can find themselves at odds. While it is impossible to describe every scenario in which conflict could arise, these examples are intended to help all team members practice awareness of how their actions and decisions can impact each other and the project.

Example: Trade-Offs between Costs and Accuracy

During its earliest stages, a development project is often a concept rather than a concrete opportunity. The developer may need to begin evaluating sites despite not yet having firm commitments from lenders or investors. This means that the developer must spend his/her own money before knowing if a project is viable and take the risk that those funds will never be recovered if the project ultimately does not come to fruition. Thus, understandably, developers can be reluctant to incur costs early on in the process. However, this frugality has consequences for the development team, particularly for the civil engineer. If a developer does not spend money to order new boundary surveys, for example, project engineers must rely on whatever public data is available. This data can be outdated and inaccurate, especially for purposes of the detailed technical analysis done by civil engineers. Using inaccurate data can lead to complications later in the development process should the project move forward, for example, if a property line was represented inaccurately. Although less than ideal, this situation is often unavoidable and the team needs to be clear about the quality and accuracy of information being used as well as any possible future implications.

Example: **Miscommunication**

Development is a complex process involving the coordination of numerous studies, submission requirements, and deadlines. Against this backdrop of commotion, even a simple miscommunication can have serious consequences. For example, specific paperwork for a permit filing may not get submitted if the developer believed the civil engineering team was handling the requirement while, simultaneously, the engineer thought the developer was making the submission. This type of situation is perhaps more common than it should be and can result from unclear expectations, a poorly defined SOW, or from simple lack of communication. Regardless of the cause, failure to submit necessary documents in a timely manner, whether to other team members or to the local Office of Planning and Zoning, can have serious effects. In a worse case, failure to make certain submissions could even mean missing important approval deadlines. Delays, especially when repeated or exacerbated, can be costly and impact overall project feasibility.

Miscommunication issues can also arise because several commonly used industry terms have different meanings to different team members. For example, the term "completion" may be understood by engineers and construction professionals to refer to the end of construction; however, a project may only be "complete" from the developer and/or broker's perspective once the building is actually leased. Different uses of the word "completion," such as "substantial completion" or "building completion" are legal terms of art and are defined in contracts, financial documents, and leases to avoid misunderstanding. However, these definitions may vary across different contracts and/or be tied to different conditions or events. Similarly, the term "commercial" may have different meanings to brokers, developers, homebuilders, and engineers depending on whether it is being used to refer to zoning, project scale, or product type. As a final example, there is no single industry standard for certain design documents, such as "concept drawings," leaving different team members to assume such plans may include different levels of detail based on their prior personal experience. To avoid confusion, a good developer and his/her team must agree to clear and specific scopes of work, communicate regularly for coordination and to reaffirm roles and responsibilities, and set up a shared project schedule to help monitor project deadlines.

Example: **Unrealistic Schedules**

Developers may sometimes ask their teams, including architects, engineers, and attorneys, to produce results in an extremely condensed timeframe. This can lead to long days (and nights) of work as civil engineers or other team members struggle to meet an unrealistic schedule. It is important for the developer to be clear about the cause of schedule compression and for the team to address it appropriately. When caused by an overzealous developer who is impatient to see results, such deadlines are obviously inappropriate and unhelpful to the team. Part of the developer's professional responsibility is to recognize that quality technical work takes time and to allow for a reasonable schedule in order to ensure the team has the time needed to create a successful project. By failing to do this, a developer risks damaging professional relationships and may create larger problems for the project if hastily completed work contains errors or assumptions that impact later project stages. However, members of the development team must understand that their work is not isolated from the rest of the project; rather, the collective work of the team's civil engineers, attorneys, architects, and lenders is all tied to interrelated deadlines and inputs in other project areas. In this context, the team should appreciate that rushed requests from the developer may be driven by a legitimate need to respond to deal pressures. For example, a developer may need cost information quickly in order to update financial projections for a loan submission deadline that was suddenly brought forward because of a rescheduled Loan Committee meeting. If the developer must wait for the next committee meeting, the project could face significant delay, which might create stresses for payment deadlines negotiated in a purchase contract or lease.

1.3. OVERVIEW OF THE DEVELOPMENT PROCESS

Development can be a long and often difficult process. Many existing efforts to describe it try to impose decision matrices or seek to frame it within discrete activities like "planning" or "execution." However, the development process is not linear and there are far too many completely unique scenarios to be funneled effectively through standardized decision-matrix gateways or into oversimplified phases. Development is a veritable tornado of activity, decisions, personalities, negotiations, and conflicts that unfold in unpredictable ways. There is no one "right way" for development to happen, even under ideal circumstances. Efforts to predict and categorize every possible scenario are essentially futile.

Rather than try to define a set process, this book identifies key development milestones that every development project must achieve and adopts the position that the "development process" is equivalent to whatever path the developer must take to reach all milestones. Specially, there are 10 significant development milestones as shown in Fig. 1.2. These are often, but not always, completed sequentially. The milestones may also be achieved concurrently, repeated, or occur out of order; however, they must all eventually be reached in ground-up development projects. The 10 development milestones that must be achieved for all projects are:

(1) recognizing opportunity, (2) selecting a site, (3) controlling the site, (4) completing due diligence, (5) acquisition, (6) obtaining approvals, (7) finalizing design, (8) securing debt financing, (9) construction, and (10) occupancy and operations or sale.

Although these 10 milestones generally do occur in chronological order, development projects are complex and the path to project completion is almost never strictly linear. It is entirely possible for some developers to reorder these milestones, depending on project circumstances. Further, successfully moving from one milestone to the next does not guarantee that a project will ultimately be successful. Developers can encounter fatal obstacles at any point, which force them to repeat previous efforts or even start the development process over from the very beginning.

It is also important to note that the activities related to the development milestones may overlap. For example, while a project is under construction (Milestone 9), the developer is already working with his/her broker to market the property for lease or sale (Milestone 10). As a result, different development team members may be involved at different times with significant overlap, although their activities may just as likely never overlap at all. The developer may have several different firms and consultants working on different parts of the project under different contracts at any given time.

All the activities occurring throughout the development process create significant leadership, project management, and coordination challenges. If not understood and managed properly, this can cause frustration or even lead to significant economic losses and project delays when work must

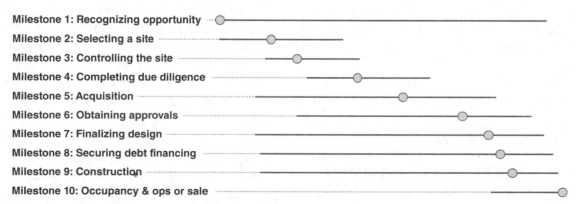

FIGURE 1.2 The 10 development milestones.

be redone or when costly mistakes are realized. It is important for civil engineers and other members of the development team to appreciate the complexity of the development process in order to better understand why the developer may sometimes seem distracted yet, at other times, be intensely focused on one specific task.

Another key point for the development team to be aware of is that the developer will have to begin spending money before he/she knows if a particular project is fully viable. Even for a large, well-funded development firm, these early expenditures represent risk because the spent funds may never be recovered. In fact, as a developer moves through the different stages of the process, his/her level of risk exposure and amount of money expended both increase. For example, attorney fees spent contracting to control a desirable site (Milestone 3) will be lost if the site fails to prove acceptable upon due diligence inspections (Milestone 4). The cost to complete the due diligence studies themselves will also be lost in this scenario.

Understanding the cost/risk element of the development process may help team members appreciate why their developer-client is reluctant to undertake certain expenditures, even though they seem imperative from a particular team member's perspective. The challenge for developers is to recognize which costs are necessary investments in success and which can possibly be forgone until a future stage in the development process. Developers do not always make the optimal choices in this regard. Civil engineers and other team members can help their developer-clients by first determining how their actions or requests might affect costs/risks and then by explaining the potential future impact of certain requests or recommendations so the developer can make a fully informed decision.

The next sections will briefly introduce each milestone in more detail and provide examples of how it may overlap or otherwise interact with other development milestones.

1.3.1. Milestone 1: Recognizing Opportunity

Exactly when and how a development project begins is difficult to define because not all projects start from the same point. For example, for a large development firm, "opportunity" may be clearly signaled by the obligation to place recently raised equity or by an internal investment committee directive. If the firm specializes in a particular product type, such as multifamily apartment development, then the developer already knows what type of project will be pursued. A specialty developer in this circumstance may need to obtain a few internal approvals but will generally spend very little time or money in order to achieve Milestone 1 and will move quickly to finding a site.

Unlike specialist developers, entrepreneurial developers often begin their process by looking for opportunities, which could mean finding either a site or a potential tenant looking for a built-to-suit solution. In many cases the entrepreneur may effectively combine the first and second milestones and look for a well-located, affordable site. In this case, identifying the site becomes the impetus for recognizing opportunity. With a site identified, the entrepreneurial developer will then define the project based on site constraints and market demand.

Occasionally, a developer will already own or control a site. For example, this may occur as a result of a previously failed project wherein the site is still under contract or may have even been purchased. In this case, "recognizing opportunity" may involve evaluating alternative uses for the site or looking for new tenants. In this instance, Milestones 2 (selecting a site) and 3 (controlling a site) essentially occur before Milestone 1 (recognizing opportunity). Depending on the circumstances, the developer may also have already completed Milestones 4 (completing due diligence) and 5 (acquisition). However, if the site was inherited through a corporate merger or as a result of a joint venture, it is possible that the developer will not have completed any due diligence and yet will have "inherited" the completion of Milestone 4. It is worth noting these types of exceptions in order to highlight the idea that development is not a linear process that can be represented by a flow chart; rather, it is a fluid and often iterative process.

Achieving Milestone 1 takes the greatest amount of time for public sector developers and private sector developers working with build-to-suit clients. Public sector developers must go through an extensive and prescribed process to determine need and, subsequently, to define the scope of a project before any further steps can be taken. This will be described more fully in Chap. 7. For the private sector developer, having a client (a future tenant or buyer) is advantageous because it eliminates the risk that the completed asset will sit vacant without rental revenue. However, it also means the developer must work closely with the client to define the project; failing to do so can lead to frustration and wasted time and money. This means the developer and his/her client must spend time at the beginning of the development process discussing and agreeing on details related to size, location, design, and cost, among other things. In some cases, a developer may even need to begin spending funds to provide the client with sample renderings or floor plans.

Even though it is presented in this section as representing the beginning point of the development process, recognizing opportunity is essentially a constant activity for private sector developers. The developer may discover a new tenant or demand trend that require changes to design considerations later in the development process. While this may be stressful to the design team, capturing such an opportunity may have significant influence on the success of the project. Similarly, even if the developer's original intention was to retain ownership of an asset, he/she may uncover an unexpected and compelling opportunity to sell the project at any point during the development process. Developers must always be aware of potential opportunities, regardless of the state of any current projects. Thus, recognizing opportunity should be understood as a continual activity.

1.3.2. Milestone 2: Selecting a Site

Selecting a site is a necessary milestone for any development because, obviously, no project can be built without a site. Thus, at some point, all developers must complete Milestone 2. The process of identifying and evaluating potential sites to determine project feasibility is called *site selection*. This includes both macro-level considerations, such as regional economic trends, and micro-level considerations, such as assessing local submarket demand for the type of development planned and evaluating competition from other, similar

projects. Developers often engage the services of a commercial real estate broker to assist in identifying sites. During site selection, it is considered a best practice for the development team, especially the architect and civil engineer, to make preliminary evaluations of potential sites identified by the developer and broker. This may require the developer to begin incurring fees even though he/she may be reluctant to do so, particularly for a site that he/she does not control. Site selection will be outlined in greater detail in Chap. 4.

Although most developers must first find a site in order to move forward with a project, a developer may sometimes already own or control a suitable site upon which to build as previously stated. In such cases, Milestone 2 is already achieved and requires little or no time and investment by the developer.

1.3.3. Milestone 3: Controlling the Site

Once a potential site has been selected, the developer must gain control of the site in order to build. The developer's broker and legal counsel are important in supporting the developer in negotiations with the site owner. While many negotiations are straightforward, others are decidedly not so and may require deal terms to be structured creatively in order to appeal to both parties and bring a deal to fruition. Note that there is no guarantee that the developer and the seller of the chosen site will reach agreement on terms, in which case the developer forfeits all current efforts to achieve Milestones 2 and 3 and returns to searching for another site or looking for another opportunity.

However, if negotiations seem promising, the developer will begin incurring legal fees to in order to negotiate either an option contract or a purchase and sale agreement (PSA). The details of these contract vehicles will be discussed in Chap. 4; however, a brief explanation is offered here. An *option contract*, as the name implies, gives the developer the unilateral right, or *option*, to purchase a particular site during a specified period of time if he/she chooses to do so. Essentially, the developer is buying the right to temporarily reserve a site. If he/she is willing to spend the money, a developer can option multiple sites simultaneously. In contrast, the PSA is an actual obligation to purchase a specific site subject to certain conditions being met.

These contracts are naturally longer and more complex than options. In the case of either contract, the developer will be obligated to make a substantial payment to the seller in the form of either an option fee or a deposit. These funds may or may not be refunded if the acquisition ultimately does not occur.

Though civil engineers are, unfortunately, rarely involved in contract negotiations, there are opportunities for them to proactively help their developer-clients by providing pertinent site information before a contract is signed. For example, a civil engineer might be able to demonstrate that the developer could achieve a more efficient layout if it were possible to acquire an adjacent parcel of land. While this would increase up-front project costs, it would also add value for the developer if it increased the overall usability of a site. The developer may (or may not) actually be able to acquire additional property, but could still use this information to potentially negotiate a lower purchase price on the subject property by demonstrating its constraints and, therefore, limited value.

1.3.4. Milestone 4: Completing Due Diligence

Due diligence is the process of analyzing a real estate asset in detail and evaluating overall project feasibility prior to acquisition. The purpose of due diligence is twofold: first, to verify there is not a prohibitive condition that might cause the developer to decide not to buy the property; and second, to discover as much as possible about existing conditions so that the developer and development team can make accurate project assumptions and determine any impacts on investment returns. In order to complete certain due diligence, the seller must agree to allow the developer and his/her consultants to access the property to conduct tests. This should be explicitly permitted in the PSA or option contract.

In the extremes, the due diligence period can range from as little as 30 days to an open-ended timeline tied to receiving certain development approvals; the former favors the seller and the latter favors the developer. In instances where due diligence is tied to approvals, Milestone 4 and Milestone 6 (obtaining approvals) overlap. In such cases Milestone 5 (acquisition) occurs *after* Milestone 6, again highlighting the point that the development milestones do not necessarily need to be achieved in a linear fashion.

The entire development team is often involved in the due diligence process and may have only a short period of time to assess the site, discover hidden challenges, and determine their impact on project feasibility. The main types of due diligence activities can be classified as technical/physical, legal, and market/financial. These will each be described in greater detail in Chap. 5, but a summary description is offered here. Technical due diligence refers to evaluating the physical conditions of a parcel in order to begin initial site design and determine the extent and cost of site work necessary to prepare the site for construction. For purposes of this book, technical due diligence also includes identifying existing environmental conditions and associated potential liabilities, usually contamination. Legal due diligence can refer to several things including a review of legal documents or contracts related to the land, land use law, or special restrictions. The results of technical, environmental, and legal due diligence can all impact decisions about the layout of a site and the design of a building. Financial due diligence pertains to developer's ability to generate acceptable project returns and secure financing for the project, both of which are tied to projected future rental revenue, or *cash flow*. Generally, development projects are only undertaken if they produce sufficient financial incentives for the developer and his/her investors. Thus, the developer will be refining his/her financial predictions based on market conditions and actively exploring any necessary debt and equity commitments during the due diligence phase.

Due diligence is necessary in order to achieve an accurate understanding of site conditions and make informed decisions about design and cost; however, completing it means the developer must spend money on tests, surveys, attorney fees, and other costs. These expenditures become sunk costs and will not be recovered if the project does not come to fruition. This represents a major risk for developers and is a key point for all members of the development team to consider.

In some cases, site conditions and other challenges discovered through due diligence inspections can be overcome. In other cases, however, the project may be deemed nonviable. If the due diligence studies reveal fatal flaws, the project will be abandoned. What exactly constitutes a "fatal flaw" depends entirely on the individual developer and his/her risk tolerance, level of sophistication, and the requirements of key debt or equity partners. For example, the need to conduct environmental remediation may constitute an acceptable risk/cost to some developers but not to others. If the results of due diligence explorations are manageable, the developer will proceed with the deal and move on to acquiring the site to complete Milestone 5.

As previously noted, development deals are complex and the development process is often not a linear one. It is entirely possible that due diligence results will lead to an unclear decision in which the developer is neither prepared to abandon the deal nor willing to move immediately to purchase the site. For example, due diligence may reveal a previously unknown condition that requires further study. In this case, the developer may need to negotiate a contract extension to further evaluate the site and will likely incur additional costs.

1.3.5. Milestone 5: Acquisition

Once a developer finds a site that satisfactorily passes due diligence inspections and can accommodate the desired project from a preliminary design standpoint, he/she will usually need to purchase it in order to move forward with construction. This is interchangeably referred to as an *acquisition*, *closing*, *purchase*, or *completion of sale*. There are circumstances in which a developer does not purchase a site outright but secures a long-term ground lease or forms a joint venture partnership with the existing property owner. For purposes of this book, any of these outcomes are equivalent to an acquisition in that they each allow the developer to have long-term use of the site for the intended project. Ground leases will be discussed in Chap. 3 and joint ventures will be discussed in Chap. 5.

With respect to a standard purchase, as with all other components of the development process, the timing of the actual acquisition is often not straightforward and may be dependent on what terms the developer negotiated in the PSA. For example, if the contract includes an approval contingency, the developer will not be required to close on the property until he/she receives all necessary approvals, or *entitlements*, from the local government to build the desired project. As previously stated, in such instances, Milestone 5 may overlap with others. If the approvals are not ultimately received, the developer will return to the site search and forfeit the time and cost spent on the project. However, not all sellers are willing to offer such a clause and the developer may have to close on the property before knowing if can actually be used as intended.

Note that the acquisition milestone is sometimes achieved concurrently with Milestone 8 (securing debt financing), depending on the sources and structure of funding for the project.

1.3.6. Milestone 6: Obtaining Approvals

Obtaining approvals, including necessary permits, must be achieved before the developer is legally allowed to move forward with construction. If the developer purchases a site that already has the necessary approvals in place, this milestone can be achieved quickly (and perhaps concurrently with several other milestones). However, for projects in many jurisdictions, this journey can be long and arduous. Approvals require the submission of detailed design plans, review by multiple government agencies, community input, and public hearings. Any stakeholder at any point in the approval process may raise objections that materially impact a developer's proposed project.

If a developer seeks approval for a permitted use, the approval process will be less complicated than if the developer seeks to change the existing zoning of a site. Zoning laws, or *ordinances*, are instituted by the authority of a local legislative body, such as a county board, city, or town council and recorded as part of the jurisdiction's code of laws. They are typically overseen by an Office of Planning and Zoning, which is responsible for evaluating potential development projects and issuing (or denying) approvals and permits. Zoning ordinances specify the use and size (or density) of buildings that are permitted on a particular site; these are referred to as *by right* uses. Some special exceptions can be granted for similar or compatible uses on a particular site, but most requests for a different use will require a zoning change. If the developer wishes to change the zoning of a particular site, he/she will need to convince the presiding Office of Planning and Zoning as to the merit of the change and such requests will be subjected to a high level of scrutiny.

Zoning will be discussed in Chap. 2 and the challenges of the approval and permitting processes, including the costs of associated fees and proffers, will be discussed in Chap. 6.

1.3.7. Milestone 7: Finalizing Design

Design is a fluid activity that evolves throughout the development process and becomes more permanent as a proposed project matures. Different components of design can change in response to physical, financial, legal, or market pressures. Further, requirements and modifications introduced during the review and approval process can materially affect the design and development program of a project. *Concept design* refers to the first stage of the design team's efforts to translate their developer-client's programmatic thoughts and requirements into a workable design solution for a particular site. The main goal of this preliminary design activity is to capture the developer's project vision and confirm that the scope can be achieved. The end product of a preliminary design exercise is a graphical/visual representation of both the proposed project and any governmental restrictions that must be accommodated on the site in order to implement the project as envisioned. *Schematic design* represents a refinement of the concept design and normally occurs after the completion of due diligence. As schematic design progresses, the developer will begin to use this information to obtain construction cost estimates in order to ensure the proposed project is achievable as originally envisioned or, if not, to identify adjustments to the design that must be made to meet project budget constraints. Although more detailed than the concept design, schematic designs are interim layouts that will serve to support subsequent, final detailed design efforts. The act of finalizing design is a necessary milestone for a development project because final design documents serve as the basis for the creation of a complete set of construction documents. *Construction documents* (CDs)

include all written and graphic documents prepared to communicate the project design and construction requirements to a building firm.

Note that while final design must eventually be achieved for all projects, under some project delivery methods it is possible to begin construction before designs are finalized. This usually applies to design-build contracts, which will be discussed in Chap. 6. In such cases, Milestone 7 (finalizing design), Milestone 8 (securing debt financing), and Milestone 9 (construction) can happen concurrently or out of order. Another example of nonlinearity can occur if a developer encounters serious financial difficulties and, shortly after starting construction, must liquidate and sell the project. In this case, the subsequent developer that purchases the project will inherit the existing final design and construction documents; assuming that he/she intends to continue with the project as-is, this second developer will achieve Milestones 5, 6, 7, 8, and 9 concurrently. Note that the new developer will have completed Milestones 1 to 4 with respect to the troubled site before acquiring it.

Final design, construction documents, construction, and delivery methods will be discussed in Chap. 6.

1.3.8. Milestone 8: Securing Debt Financing

A core tenant of development is the use of "other people's money." This refers to the combination of debt and equity sources a developer brings together in order to finance a project. The total financial capital used to finance a project is referred to as the project's *capital stack* and includes all sources of debt (from lenders) and equity (from equity partners and/or the developer). The equity portion of the capital stack will almost always be significantly less than debt portion. Thus, it is a necessary milestone for developers to secure debt financing.

There are two types of loans developers must typically obtain in order to bring a project to fruition: construction loans and permanent loans. *Construction loans* are interest-only loans that function similarly to a line of credit wherein the developer draws funds on an as-needed basis to cover the ongoing cost of site work and construction. Construction loans are usually paid off by the placement of the permanent loan, although some construction loans include terms that allow them to convert to permanent loans. *Permanent loans* have conventional financing terms and are placed when the finished building reaches *stabilization*, the point at which cash flow from rental income is sufficient to support operations and generate some minimum level of positive net profits. Given that these two types of loans are needed at different times during the development process, rather than occurring as a discrete event, Milestone 8 spans an extended period of time and overlaps with other development milestones. Developers need to begin evaluating their options for debt as early as Milestone 1 in order to have a realistic idea of what type and scope of project to pursue. Considerations for the type and timing of debt placement will be described throughout Chaps. 3 to 6.

Developers are not required to include either lenders or equity partners as sources of funds in a project's capital stack if they are able and willing to contribute the necessary equity themselves. However, private sector developers rarely complete projects on an all-cash basis; the cost and risk to do so are prohibitive. In contrast, most public sector developers do not use bank loans or other traditional debt financing mechanisms. Rather, the public sector undertakes development projects on an all-cash basis financed either through tax revenue or the issuance of bonds. Thus, this is the only milestone that does not typically apply to public sector developers. Note that certain forms of intragovernmental revolving funds and grants do exist, usually for large highway infrastructure projects, but these are not sourced from traditional lenders.

1.3.9. Milestone 9: Construction

Construction is a symbolic milestone because, for the first time in the development process, there is visible evidence of a project being built. There are two ways to classify construction: as *horizontal construction*, which refers to *site work* or physical changes made to the land, and as *vertical construction*, which involves the actual construction of the building itself. During construction, the developer's focus is on maintaining the construction schedule and moving through the necessary inspections efficiently; delays or changes during this stage will increase costs and reduce project

returns. Additionally, long before construction is completed, the developer must already begin marketing the property for sale or lease. These efforts, if successful, can improve project returns by reducing the time to stabilization. Also note that for some projects, construction can continue even as parts of the project are beginning to be occupied. This is true when final interior work is being done on different floors of the same building, across different phases of larger projects, or as homes are built in different phases of a residential development. Thus, construction can often overlap with Milestone 10 (occupancy and operations or sale).

There can be some ambiguity with respect to determining when the construction portion of a project is actually completed. Measures such as contractually defined *substantial completion* or receipt of a Certificate of Occupancy from the local government can be used. A developer may also define the end of the construction period as the date on which the construction loan is retired, which can occur long after the receipt of a Certificate of Occupancy. These and other components of construction will be discussed in Chap. 6.

1.3.10. Milestone 10: Occupancy and Operations or Sale

Although many members of the development team have concluded their roles once construction is completed, the development project itself is not finished. A completed building that is *empty* does not represent a successful development. The building must be leased or sold in order to create economic value and repay the debt that was used to finance the project and pay development team members' fees.

A building is said to have been *delivered* to the market when it is ready for tenants to take occupancy. Only at this point is the developer finally in a position to start recognizing cash flow from the development project. Depending on the success of marketing activities during construction, the developer may already have a significant number of signed leases and tenants waiting to move into the building. The number of tenants and specific terms of their leases will determine when the property reaches stabilization and cash flow from rental income is sufficient to support operations and generate net profit. If the developer is a merchant builder, it may still be necessary to go through the leasing process in order to create an attractive rental revenue stream for an investor. If the developer intends to retain ownership of the building, he/she will now be beholden to the results of prior decisions regarding building systems, layout, amenities, and countless other factors will impact long-term leasing and the cost of operations.

It should be noted that this milestone is slightly different for public sector developers, who are not seeking revenue from their newly completed projects. However, they will still occupy and operate (or otherwise use) the project and do not achieve any value from the development until they do so.

CHAPTER CONCLUSION

This chapter described the typical characteristics of developers and examined the roles and responsibilities of key members of the development team. Examples of conflict between the developer and development team members were presented. The chapter also provided an overview of critical milestones in the development process. The next chapter will describe the complexities of the built environment, including zoning, in which developers and their teams operate.

REFERENCES

1. Bureau of Labor Statistics, U.S. Department of Labor, *Occupational Outlook Handbook, 2016–17 Edition*, Civil Engineers, available at https://www.bls.gov/ooh/architecture-and-engineering/civil-engineers.htm (accessed August 26, 2017).
2. NCEES PE Civil Exam information, available at http://ncees.org/engineering/pe/civil/ (accessed August 26, 2017).
3. National Council of Architectural Registration Boards website https://www.ncarb.org/become-architect/basics (accessed August 19, 2017); AIA website https://www.aia.org/pages/2651-getting-licensed (accessed August 26, 2017); NCARB Architectural Experience Program Guidelines brochure 2017 available at https://www.ncarb.org/sites/default/files/AXP-Guidelines.pdf (accessed August 26, 2017).
4. AIA document library: https://www.aiacontracts.org/ (accessed August 26, 2017).
5. The Law School Accreditation Process, Revised September 2016, American Bar Association Section of Legal Education and Admissions to the Bar, Chicago, IL, available at URL: https://www.americanbar.org/content/dam/aba/publications/misc/legal_education/2016_accreditation_brochure_final.authcheckdam.pdf (accessed August 26, 2017).

CHAPTER 2

CREATING THE BUILT ENVIRONMENT

The previous chapter introduced different types of developers, their development teams, and the development process. This chapter will describe the physical and regulatory environment in which developers and their teams operate. Although the development milestones introduced in Chap. 1 will not be discussed directly, the information presented herein provides a critical context that applies to all projects. No development project happens in isolation, and it is incumbent upon the developer to understand and navigate the constraints placed on the development process by different components of the surrounding environment.

2.1. OVERVIEW OF THE BUILT ENVIRONMENT

Whether you live in a large city or a rural community, everything you see in the world around you is part of the *built environment*. The built environment is a term that refers collectively to the complex web of structures, roads, utilities frameworks, and communication pathways that make up the physical world we live in. The built environment includes *everything*. It includes things that most people see or think about every day, like houses, schools, offices, and stores. It also includes things that people are less actively aware of: warehouses, monuments, roads, bridges, mines, power plants, electrical lines, and water treatment plants. The built environment also includes specialty structures, such as sports stadiums, hospitals, museums, airports, prisons, fisheries, and racetracks. Even open spaces, parks, cemeteries, and natural reserves are part of the built environment because, although they are not necessary man-made, deliberate decisions were made to *not* develop these parcels and to preserve them so they can contribute to the quality of the human experience. Arguably, the only parts of the world not included in the built environment are large bodies of water—and even these have been partially developed with dams, docks, ports, harbors, and commercial shipping facilities. Given that the built environment is so vast, it is often divided into function-based subcategories, making it easier to visualize and discuss. The main categories are (1) infrastructure, (2) open/green spaces, and (3) buildings/structures.

Infrastructure refers to the components of the built environment that support a nation's economic and societal ability to function. There are several different kinds of infrastructure. For example, interstate highways, roads, bridges, subway systems, and train lines are referred to as part of the nation's *transportation infrastructure*. Without transportation infrastructure, it would essentially be impossible to get people or products from one place to another. Similarly, water and sewer systems, telecommunication systems, and gas or electrical lines are part of the *utility infrastructure*; without these components of the built environment, modern life as we know it would not be possible. Infrastructure can also be subcategorized as either hard or soft. *Hard infrastructure* refers to physical systems (roads, power lines, sewer systems, etc.) and *soft infrastructure* refers to the institutions that support society, such as schools, the police force, and fire departments. Sometimes these same categories are also referred to as *physical infrastructure* and *social infrastructure*. Hard (or physical) infrastructure projects are known as *capital improvements*.

While infrastructure supports the built environment and facilitates human activity, open spaces and green spaces enhance the human experience and help preserve the environment. Most *open/green spaces* are available to the public and generally feature minimal or no structures and little infrastructure; examples include parks, playgrounds, municipal sports fields, and public plazas. The nation's extensive system of national parks is also included in this category, meaning that open/green spaces encompass more total square footage than either of the other two categories of the built environment. There is a subtle difference between "open" spaces versus "green" spaces based on how much of a particular site is covered with grass and natural vegetation; for purposes of this book, the distinction is not important and all related uses will be referred to collectively as "open/green" spaces unless otherwise specified.

The largest category of the built environment in terms of investment is buildings and structures. *Buildings/structures* essentially refers to all built structures that are not related to infrastructure. It encompasses everything from small, privately owned family homes to office skyscrapers owned by large corporations. The government can also own and develop buildings/structures, such as courthouses or police stations. Of the three categories of the built environment, buildings/structures is the most diverse because buildings are designed and used for a wide array of purposes: office, residential, retail, warehouse, industrial, military, and specialty uses. Further, this component of the built environment also includes things such as watchtowers, monuments, and statues.

Regardless of category, no component of the built environment appeared suddenly or arbitrarily; decisions were made, often years in advance, about what would be built and where. Sites had to be identified and purchased, which requires complex legal instruments and contracts. In the case of private sector development, extensive research was conducted to determine viable uses. Further, all elements of the built environment had to be paid for, also known as being *financed* or *funded*, before their construction could begin. And finally, virtually all components of the built environment had to be constructed. As explained in Chap. 1, the process of finding a site, determining what should be built on it, obtaining funding for the project, getting permission to build, and overseeing construction through to completion, is known as development. Thus, the built environment is created through the development process.

However, developers and their teams do not create the built environment in isolation. Rather, multiple different actors are involved in different stages and in different capacities. These can be divided into two categories: the public sector and the private sector. As stated in Chap. 1, the public sector refers to the government and its agencies and departments; these can be federal, state, or local. Collectively, the public sector manages and enforces a complex system of laws and processes that govern how the built environment is created. The public sector is also involved in the development, ownership, and maintenance of the country's infrastructure networks and publicly owned buildings and structures. The private sector is also a key actor involved in creating the built environment. In its broadest sense, the private sector is essentially any nongovernmental individual or entity. One characteristic that distinguishes private sector actors from their public sector counterparts is the private sector's sensitivity to risk, timing, and investment returns.

The next sections explore the different components of the built environment in greater detail. They will also explain how the public sector attempts to guide and manage the development of the built environment through a process known as planning.

2.2. COMPONENTS OF THE BUILT ENVIRONMENT

The built environment is large and complex. Given that it is impossible for any development project to occur in isolation, it is important to understand the context surrounding new projects. There can be costly consequences if a developer fails to account for obligations imposed by different components of the built environment. In a worst-case scenario, failure to comply with the requirements of the built environment can easily jeopardize a project's viability and harm the developer's credit and/or reputation. This is important not only for developers to understand but also for civil engineers and other members of the development team. The more thoroughly the team understands how these elements impact their developer-clients, the better support they will be able to offer.

This section will examine individual components of the built environment in greater detail in order to help clarify how each can impact a development project. Specifically, this section will consider (1) infrastructure, (2) open/green spaces, and (3) buildings/structures.

2.2.1. Infrastructure

Infrastructure is comprised of those elements of the built environment that facilitate a nation's societal and economic activities. Because it is a large and diverse category of the built environment, there are many competing theories with respect to what *exactly* should be included in the definition of infrastructure. This book is primarily concerned with the types of infrastructure that are most commonly and directly connected to development projects. This includes (1) water and sewer systems, (2) transportation, specifically roads and highways, (3) electricity and gas, and (4) select forms of soft (or social) infrastructure, such as schools.

The U.S. population has continually increased since much of our core infrastructure was first built and, naturally, an expanding population requires greater usage of infrastructure services: there are more cars on the roads, more homes using electricity and water, and more children in schools. This, in turn, leads to a rise in wear and tear on existing infrastructure and, often, the need to increase the capacity of the underlying systems. However, investment in infrastructure improvements has not kept pace with society's needs. The American Society of Civil Engineers (ASCE) evaluates the state of U.S. infrastructure systems every four years and issues an "infrastructure report card" based on existing conditions. The results in recent years are lamentable. U.S. infrastructure received an overall grade across all systems of D in 2009, D+ in 2013, and again D+ in 2017.[1] The estimated cost to bring our infrastructure up to an acceptable standard is hundreds of billions of dollars, and the ASCE estimates there is as much as a $2 trillion funding gap over the next 10 years.[2]

The state of our infrastructure is relevant to development for two reasons. First, because it affects the costs of virtually all projects and, second, because the combination of existing infrastructure conditions and lack of funding have led to an opportunity for the private sector to become increasingly involved in the development of infrastructure projects through partnership with the public sector. The next sections of this chapter are dedicated to a discussion of the infrastructure systems themselves and their impact on development costs.

Water. Society's need for water is enormous. In 2016, residents of New York City used an average of 117 gallons per person per day, which translates to well over 1 billion gallons of water used each day by the city's population.[3] The home is not the only place where people use water: there are bathrooms and break-rooms in every office building and school, kitchens in all restaurants, showers and water fountains in gyms, and countless other taps of different kinds. The U.S. Army Corps of Engineers Institute for Water Resources estimates that "it takes nearly 1,200 gallons of water per person per day to meet the total needs of a city including schools, factories, offices and businesses and the many other private and governmental organizations that run a city and make it possible [to live] our daily lives."[4]

This massive demand is met by an extensive infrastructure system. Water is collected directly from rivers, groundwater sources, and also through the use of dams. There are more than 5,500 dams across the United States, the primary purpose of which is tied to providing adequate water supply.[5]

Most of these are managed by federal, state, or local entities. Although they support water collection, dams themselves are not uniformly considered to be part of the water infrastructure system. Rather, dam structures are often classified as their own, unique type of infrastructure.

Water supply is not the only source of society's water-related needs: unwanted wastewater and excess stormwater must be removed and treated appropriately. This is handled through an equally extensive pipe network that routes water to treatment plants or reservoirs. Water supply, wastewater removal, and stormwater management must all be considered in development projects.

Water Supply. The *water infrastructure* system refers to components that collect, purify, and distribute water for human consumption through a series of reservoirs, tanks, pumping stations, and pipes. This includes *potable water*, which is safe to drink, and nonpotable water, which is used for manufacturing and other purposes. Distribution of potable water is usually handled by a combination of entities, starting with a water "wholesaler" who supplies clean water to smaller utility firms or municipalities who, in turn, deliver the water to individual customers at their homes or businesses.

The U.S. Army Corps of Engineers operates and maintains 380 water reservoir projects across the United States.[4] While not all of these reservoirs are used to supply drinking water, many are. In the Washington, D.C. area, the Corps provides drinking water by collecting, purifying, and distributing water from the Potomac River. This work is done through a federally owned entity known as the Washington Aqueduct, which was created by an Act of Congress in 1859 as a public water supply agency to provide a supply of drinking water for D.C. and Northern Virginia.[6] The Aqueduct's processing facilities include the Dalecarlia Water Treatment Plant and the McMillan Water Treatment Plant, which were built by the U.S. Army Corps of Engineers and have been in service since 1863.[6] Collectively, these facilities produce "drinking water for approximately one million citizens living, working, or visiting the District of Columbia, Arlington County,

Virginia, and the City of Falls Church, Virginia" and the surrounding area.[7]

Though it serves the area, the U.S. Army Corps of Engineers does not provide water directly to citizens or corporate users. Rather, the Washington Aqueduct is a "potable water wholesaler," which sells the water it produces to local governments and municipalities. The local governments then deliver the water to individual homes and businesses in their jurisdictions in exchange for a fee paid by the homeowner (or business owner) in the form of a utility bill. This is a common arrangement for water supply in the United States.

Part of the cost of water utility bills is based on the need to maintain the extensive infrastructure system that delivers water from the water wholesaler to local governments (or other users). It costs more than $40 million each year to operate and maintain the Washington Aqueduct's physical infrastructure plus an additional cost to develop capital reserves for future improvements and major repairs.[6] This cost is paid by the Aqueduct's wholesale clients: the local governments, who must then build the cost into their water charges to citizens.

Local jurisdictions have their own share of infrastructure to maintain in order to transport water from a source or wholesaler to their citizens' individual homes and business locations. Of the Aqueduct's service areas, Arlington County in Virginia owns 500 miles of pipelines that distribute approximately 23 million gallons of water each day through a system of "three water storage facilities, five pumping stations, and 12 pressure monitoring sites"[8] operated by the Arlington County Water Control Center. Counties and/or municipalities, including Arlington County, must support the maintenance of their portions of the water infrastructure through taxes and/or water fees to citizens or else contract with a private sector operator, who will make similar charges to users.

Water supply in the United States is governed by an array of regulations. Some of the primary sources of control include the Water Supply Act of 1958, the Safe Drinking Water Act of 1974, and the Water Resources Development Act of 1986, as well as the National Primary Drinking Water Regulations in the Code of Federal Regulations (40 CFR 141). Of these, the Safe Drinking Water

Act (SDWA) is the primary law established with the aim of protecting the quality of drinking water in the United States. It is incorporated into the U.S. Code as 42 USC Subchapter XII—Safety of Public Water Systems. The SDWA focuses on all waters actually or potentially designated for drinking use, whether from surface or groundwater sources. The act authorized the Environmental Protection Agency (EPA) to establish safe standards of water purity, known as *maximum contaminant levels*, which are now required in all water distributed for drinking. The SDWA has been amended numerous times, resulting in increasingly stringent drinking water regulations for both municipal and small, localized on-site drinking water systems.

Water supply represents only half of our water-related needs, as represented in Fig. 2.1. Sanitary sewer systems collect "used" wastewater and return it, via an extensive pipe network, either to a treatment plant or a reservoir. The next section will discuss sanitary sewer systems and stormwater systems.

Sanitary Sewer and Stormwater.   Sewer systems, also referred to as wastewater systems, are necessary complements to water systems. Much like water systems, *sewer infrastructure* consists of pipes, pumping stations, and collection and treatment reservoirs. Unlike water systems, in which the federal government may participate in the system as a water wholesaler, public sewer systems are almost exclusively owned and maintained at the state or local level. The federal government's involvement is limited to regulation and providing

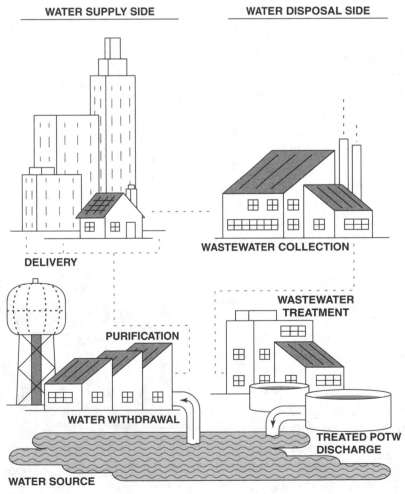

FIGURE 2.1   The water cycle.

special funding programs.[9] Though there are many different kinds of wastewater management and sewer systems, this section is focused only on centralized sanitary sewer and storm sewer infrastructure; individual, on-site wastewater systems, such as septic systems, are not included in this discussion.

Sanitary sewer systems and stormwater sewer systems serve different purposes and, in modern design, usually maintain separate infrastructure (although many older systems combine them). The storm sewer system collects rain and other external run off or drainage, while sanitary sewer systems collect wastewater from buildings and structures and transfer it to a treatment plant. Treatment plants are often referred to as "works." When owned by a government agency, they are called *Publicly Owned Treatment Works* or POTWs. There are more than 16,000 operating POTWs, which serve approximately 75 percent of the U.S. population's wastewater treatment needs.[10] There is also a small but increasing number of privately owned and operated treatment works. Many existing plants have been privatized since the 1992 issuance of Presidential Executive Order 12803 on Infrastructure Privatization allowed that, in order for "the private sector to provide for infrastructure modernization and expansion, State and local governments should have greater freedom to privatize infrastructure assets."[11]

Whether public or private, treatment works collect wastewater that contains physical, biological, and chemical contaminants that must be removed or treated before the water can be discharged into the environment or reused. This is achieved through multiple treatment phases that apply extensive physical, biological, and chemical treatments to the wastewater.[12]

The primary federal regulations related to wastewater processing and discharge are the Clean Water Act of 1948, the Clean Air Act of 1970, and the Resource and Recovery Conservation Act of 1976. The objective of the Clean Water Act (CWA) is to restore and maintain the chemical, physical, and biological integrity of our nation's water. One way the CWA achieves this is through the National Pollution Discharge Elimination System (NPDES), which limits the amount of pollutants permitted to be discharged back into the natural water supply.[13] Specifically, it establishes a total maximum daily load (or amount) of pollutants allowed to be discharged from any point source. *Point sources* are discrete conveyances such as pipes or man-made ditches. The act also addresses water supply and water pollution issues through non-point-source controls and effective use of water resource management practices. Another CWA protection dictates that sludge resulting from wastewater treatment under CWA must be handled at a Resource Conservation and Recovery Act (RCRA) facility if it is hazardous. Discharges from an RCRA-permitted facility must also be pursuant to a National Pollution Discharge Elimination System permit from the EPA. The permit provides limits on what is being discharged, monitoring and reporting requirements, and other provisions. Generally, the NPDES permit program is administered by authorized states. Construction activities are considered an industrial activity and, as such, will typically require an NPDES permit.

Implications for Development.   Imagine the vast network of water and sewer pipes hidden beneath the ground. These existing pipes are part of the utility's *capital improvements*. The pipes connected at source reservoirs or storage tanks must be able to handle bulk volumes of water and are large enough to walk through, such as those shown in Fig. 2.2. Meanwhile, the pipes connected to homes, which only need to manage the flow of a few appliances and taps, are small in comparison.

FIGURE 2.2   Water pipes.

In order for water to be available in homes, offices, or any other structure, developers must connect their new buildings to the pipes somewhere along the water and sewer infrastructure systems. In exchange for allowing developers to access the water supply, municipalities charges *tap fees*. Tap fees are charges developers must pay to cover the cost of physically creating access, or "tapping" into, the existing water and sewer lines. Developers may have to pay an *impact fee*, which is a separate charge to offset the cost of upgrading the system in order to accommodate the increase in services needed to supply the occupants of the new buildings.

The cost to bring water to a site depends on the volume of water required and the size of the pipes needed as well as the distance the pipes must be run. If a capital improvement (existing pipe) is available at the development project site, new pipes will only need to be run for a short distance and the cost may be only a few thousand dollars. However, if the nearest capital improvements are far away from the site or on the other side of an adjacent property, the cost can surpass a million dollar threshold. An adjacent site owner may be unwilling to allow a developer to run connection pipes through his/her land, leading to the need to either buy access or reroute pipes around the adjacent property. Costs can also be high in rural areas where new pipes may have to be laid for as much as a mile to connect with the nearest capital improvement. Similarly, costs associated with wastewater access can also range from a few thousand dollars per unit to several hundred thousand dollars per unit.

Unanticipated site conditions can also contribute to the cost of water and sanitary sewer connections and this is a potential area of conflict between the developer and civil engineer. A civil engineering firm will estimate the cost of work based on studies done during the due diligence phase, which will be discussed in greater detail in Chap. 5. On the basis of these studies, the engineer and developer will agree on a line item cost for site and utility work, which the developer will factor into his/her financial analysis of the project. However, when the civil engineering firm actually begins doing work on the site, such as excavating,

unexpected conditions may be discovered. These can materially change the work required and dramatically impact cost. For example, if an underground pocket of either hard rock or particularly soft soil is discovered during site work, utility pipes cannot be run as planned. They will either have to be extended and run around the problem area or the limiting condition will have to be ameliorated. Extending pipe length increases cost as does blasting rock or hauling in new soil to replace structurally unsuitable soft soil. While the engineer is concerned with solving the technical problem, the developer is concerned about project viability. If the developer has not included a contingency for unexpected or "extraordinary costs" in his/her financial projections, these unexpected increases will reduce expected project returns. If returns fall below investor or lender prescribed thresholds, the project may not go forward.

New development is also affected by the need to comply with the various federal water and wetland regulations. As previously stated, construction activities can trigger the need for an NPDES permit and a requirement to follow certain guidelines. Specifically, an NPDES permit is required if the construction results in a disturbance of more than one acre of total land area.[14] Some jurisdictions may have additional requirements based on regional or local environmental concerns. For example, in the Chesapeake Bay watershed, 2,500 square feet of disturbance can trigger the requirement for a permit. Construction activities that require a permit may include clearing and grubbing, grading, and excavation. These activities are typically associated with road building, borrow pit excavation, residential housing construction, office building construction, light industrial construction, or facility demolition. In some states, a stormwater permit may not be required if the runoff does not discharge into a waterway, as in the case where it evaporates from a catch basin or similar isolated water body. Additionally, as a result of the NPDES permit program, construction sites are required to have a Stormwater Pollution Prevention Plan prior to commencement of any land-disturbing activities on one or more acres.

As a percentage of total project cost, water, sewer, and stormwater expenses can account for a significant portion of the site development costs of a project. This may impact project affordability, both for the developer and for the end user, to whom costs are passed through in the form of either rent or purchase price. For example, recent changes to the aforementioned Chesapeake Bay Preservation Act have pushed stormwater management costs in certain situations to over $1 million per acre.

Transportation. The daily routines of both people and businesses require physical movement from one place to another. People need to drop their children off at school, get to their offices, and go to the grocery store; on weekends they might take their kids to soccer practice, go out for dinner and a movie, or travel to visit family. For many people, these and other daily commutes are made with cars, while for others such activities may involve a bus or subway system. Businesses also need to move people and products on a regular basis: they send employees on business trips, ship goods to customers, and may receive materials to be used in the production of other goods.

The components of the built environment that allow people (or things) to move from one place to another are part of the country's *transportation infrastructure*. This includes a vast system of roads, including interchanges as shown in Fig. 2.3, bridges, tunnels, railways, runways, and subways. This component of the built environment is so critical to the American economy and way of life that Congress maintains a dedicated Transportation and Infrastructure Committee (T&IC) to provide oversight to these infrastructure projects and related federal agencies.[15]

One of the agencies that falls under the T&IC's purview is the U.S. Department of Transportation (DoT). This Executive Branch agency was established by an Act of Congress in 1966 and is "responsible for ensuring the movement of people and goods throughout the United States as well as to and from our Nation's borders."[16] Among its many departments, the DoT includes the Federal Aviation Administration (FAA), the Federal Railway Administration (FRA), the Federal Transit Administration (FTA), and the Federal Highway Administration (FHWA).

FIGURE 2.3 Road infrastructure.

As is evident from the DoT's range of specialty administrations, transportation infrastructure is such a large and diverse component of the built environment that it is often broken into subcategories based on specialized modes of transportation. Thus, for example, while both roads and railways are both parts of the transportation infrastructure, they are often dealt with independently. Of all the different forms of transportation infrastructure, this book will focus exclusively on roads and highways as these are directly and physically connected to, or integrated with, development projects.

Roads Systems. America's road network is extensive. It includes everything from large interstates to small neighborhood roads as well as bridges and roads running through national park land. Public roads in the United States stretch for almost 4.2 million miles and provide American drivers with 8.8 million miles of total travel area, accounting for multilane roads.[17] An important component of the road network is the National Highway System (NHS). The NHS is a specific network of approximately 160,000 total miles of road comprised of interstates, major highways, primary arterial routes, and key connector routes in the United States.[18] While the NHS represents a small percentage of total U.S. roadways, NHS roads carry the heaviest volume of traffic and represent the most critical transportation linkages for American society, economy, and defense. An estimated 55 percent

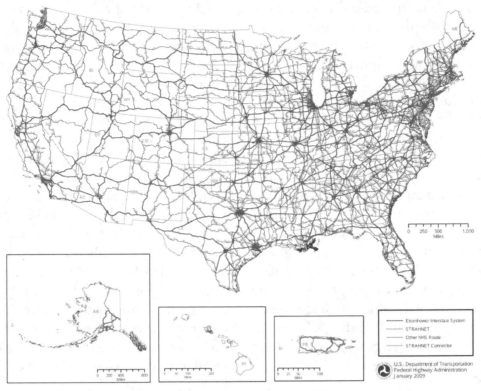

FIGURE 2.4 The national highway system. (*Source: FHWA.*)

of all vehicle travel occurs on the NHS, including about 83 percent of all truck travel.[19] Figure 2.4 shows the national coverage of the NHS.

Over the century since the introduction of motor vehicle travel, America's roadways have aged, and traffic volume has steadily increased. In 1920, during the road network's infancy, Americans drove for a combined total of 47.6 billion miles, a measure referred to as *total vehicle miles* (TVM).[17] By 2015, the TVM traveled had increased to 3.1 trillion, an increase of over 6,000 percent.[17] Over the same period of time, 1920 to 2015, increases to the capacity of the NHS trailed dramatically compared to the increases in its usage.[17] In addition to capacity issues, such a dramatic increase in usage also has implications for wear and tear and maintenance needs. In many ways, the cost of poor road conditions and capacity is borne by citizens in the forms of safety hazards, increased vehicle maintenance costs, and time lost waiting in traffic congestion. The challenge of curing these shortcomings as well as protecting air quality and other road-related concerns falls to the various federal, state, and local government agencies connected to the road infrastructure system.

Construction and maintenance of America's roads, including the National Highway System (NHS) and its interstates, is the responsibility of individual states and *not* the federal government. In fact, the federal government owns and maintains very few roads relative to the nation's total lane miles, most of which are located on federally owned land such as national parks or military bases.[20] Roadways that fall inside state borders are the responsibility of that state. However, states are required to comply with federal regulations pertaining to the design, bidding, and construction of roads in order to receive federal funding, which is available for the construction of NHS road projects. The need to provide federal funding for roads can be traced back to political discussions in the late 1800s and early 1900s connected to the federal government's interest in creating reliable "post roads" to facilitate mail delivery.[21] States have been eligible to receive federal support for road projects since the passing of the Federal-Aid Road Act in 1916.[22] Thus, though the states own, build, and maintain our roads, the federal government contributes significant funding for NHS projects.

The DoT's FHWA is the federal agency most directly involved with the management of the NHS. The FHWA provides research, regulation, guidance, and funding to "support State and local governments in the design, construction, and maintenance of the Nation's highway system."[23] The agency is integrally involved in everything from highway design standards, managing interstate access policies, overseeing federal aid for highway projects, and ensuring state compliance with federal procurement requirements for highway construction contracts.

In particular, the FHWA manages several specialized programs, such as the National Highway Performance Program and the Highway Safety Improvement Program. These programs provide guidance to states and help align eligible projects with governing regulation and funding sources. As an example, the National Highway Performance Program's (NHPP) stated purpose is[24]:

1. To provide support for the condition and performance of the National Highway System (NHS).

2. To provide support for the construction of new facilities on the NHS.

3. To ensure that investments of federal-aid funds in highway construction are directed to support progress toward the achievement of performance targets established in a state's asset management plan for the NHS.

Costs to maintain the roadway transportation infrastructure stem from both the need to increase capacity and the need to maintain or repair existing roads. Roads can be costly to build and repair, depending on conditions such as local labor costs, topography, the need to acquire land for access points, and the number of lanes (or other specifications). Roads in urban environments can be especially costly because of the need to design and build complicated intersections or access points, lack of space for workers and equipment, and a higher incidence of existing utilities that have to be accounted for and may need to be moved.

As previously stated, the federal government provides funding assistance to states for eligible roads that are part of the NHS; however, federal funding does not cover 100 percent of total project costs.[25] The source of federal highway infrastructure funding is the Highway Trust Fund. The Highway Trust Fund was established by the Federal-Aid Highway Act of 1956 as a means of ensuring a dedicated source of funding for highway projects in the United States.[26] The fund's primary source of revenue is the collection of special-use taxes paid by American drivers in the form of fuel taxes. Currently, the U.S. government collects a federal fuel tax of 18.4 cents per gallon on all gasoline purchases and of 24.4 cents per gallon on diesel fuel purchases made in the United States.[27] The higher rates charged on diesel are intended to reflect the greater damage from wear and tear inflicted by heavy and commercial trucks.[28] Most of this federal fuel tax revenue, along with revenue from other federal taxes levied on heavy trucks, is allocated to the Highway Trust Fund to be disbursed by the FHWA to fund future highway projects.

However, the Highway Trust Fund's solvency is in question. The total number of drivers and associated strain on the system has increased significantly since the fund's inception in the 1950s, which in turn increases the need for (and cost of) infrastructure improvements and capacity increases. Further, portions of the revenue collected from the federal fuel tax are diverted to non-highway-related projects, specifically general Mass Transit projects and those related to leaking underground storage tanks. Despite significant additional infusions of funds provided by special legislation, Congregational Budget Office projections show that the Highway Trust Fund will be overdrawn as early as 2021.[29] A range of solutions have been proposed to address this inevitable shortfall including increased taxes, greater use of direct user fees (tolls), and incentivizing more private sector involvement through public-private partnerships. No definitive solution has yet been realized.

Implications for Development. It is important to restate that the NHS represents only about 5 percent of the total highway miles in the United States.[30] It is the sole responsibility of states and local jurisdictions to fund, construct, and maintain the remaining (approximately) 95 percent of roads, which are not eligible for federal funding. Therefore, changes that impact local road infrastructure are very important to municipalities. This affects new development projects in two main ways: approval and access.

Depending on the size of the project, the process of obtaining approval for a new development often requires the developer to pay for a *traffic study*. There are different types of traffic studies but, generally, the main purpose of these analyses is to evaluate existing traffic conditions surrounding a site in order to determine what impact the new development will have on the road system. Traffic studies typically include a count of *trips*, or the number of cars passing by, as well as observations about traffic conditions, congestion, and accidents. The studies will usually make traffic projections for the roads or intersections included in the study area, both including and excluding the proposed development. Finally, traffic reports often make recommendations for infrastructure improvements, such as the need to widen roads or install new traffic lights. This information helps the local government understand the impact a proposed development project will have on their jurisdiction.

It is possible that a government would deny a development if the results of its traffic study were extreme. However, most often, the developer is charged a *transportation impact fee*. Similarly to water/sewer impact fees, this fee is meant to help the state or local government offset the cost of the impact the proposed development will have on the existing road infrastructure.

Understandably, developers tend to be concerned with the success of their individual projects more than the government's road systems, even though connecting to a functional roadway system can be essential for the success of a project. From this perspective, traffic studies can seem like an unnecessary cost and administrative burden. But from the municipal perspective, transportation infrastructure must be considered holistically across an entire jurisdiction. For example, if a developer proposes a new 100-unit apartment building project, adjacent roads may have to accommodate as many as 600 new cars after the project is completed due to trips by new residents, visitors,

deliveries, and employees at the building. Various professional organizations, such as the Institute of Transportation Engineers (ITE), provide resources used in estimating the exact future impact. Now imagine not just the impact of a single 100-unit project, but the impact of all the new projects built in a municipal jurisdiction in a single year, including all new office buildings, retail stores, and residential uses. Each new building brings new cars, delivery trucks, and even busses onto the municipality's road system. In some instances, this traffic is merely relocated from other areas, but in any growing town or city, the aggregate affect is new net growth and an overall increase in the use of the road infrastructure, especially when new development increases density.

Traffic studies add to the expense of a development project because the developer must pay for the study and the resulting impact fees. Further, he/she must also wait for the study to be completed and for the government to review it, which takes time and can impact the development schedule. The time during which these studies are authorized to be conducted can also be a factor. For example, most jurisdictions do not allow traffic counts to be taken during periods when schools are not in session, which can cause extensive delays for projects occurring in the summer months. Yet traffic studies are not the only way that the road system impacts new development.

A second way development projects are affected by the transportation infrastructure is related to access. All new developments must connect to the existing road system in order to allow vehicular traffic to access the new building(s). This can be a simple *curb cut* allowing street traffic to access a parking lot, garage, or loading dock. Or it can be an expensive requirement in which the developer must fund or construct off-site infrastructure improvements related to the new property. Adding a new deceleration turn lane to a street adjacent to the new development site is an example of an off-site transportation infrastructure improvement that might be required of a developer, even though the road and the new lane belong to the government. Similarly, requiring a developer to create a new intersection and install traffic lights at the entrance to his/her property is another example of an off-site improvement related to transportation infrastructure. While making such improvements can be costly in its own right, there can also be hidden costs if the developer's required work necessitates other modifications, such as widening an existing road. If existing utility lines are discovered under or near the existing road, they can be expensive to move. In some instances, the utility firms themselves may be obligated to move their infrastructure, but this can be a slow process and delay the project.

Some developers may also need to pay for on-site improvements, which are most common in large residential developments where new roads must be created to connect single-family homes and allow residents to drive through the community to reach existing public roads. These community roads are built and paid for by the developer and then *dedicated*, or given, to the local municipality, which will maintain them.

As a percentage of total project cost, road- and access-related expenses are dependent on the type of development. Costs for a new, urban office building will be different from a suburban retail shopping center or a large new residential community. The results and recommendations from traffic studies and other transportation related expenses must be included in early cost estimates during a project's feasibility phase.

Electricity/Gas. The electricity used in American homes and business is provided through a distribution system referred to as the national electric grid. In fact, the electric grid is actually made up of three separate grids, each of which include a system of power plants, high-voltage transmission lines, transformers, and substations.[31] One of these three grids, called the *Eastern Interconnection*, also provides power to part of Canada. From substations, electricity is transferred to buildings through a network of smaller, local transmission lines to end users. The power plants on the grid system use various primary energy sources to generate electricity, including coal, natural gas, hydropower, nuclear, wind, and solar energy. The electricity from all of these sources feeds into the same grid system.

Because electricity travels so quickly, it is produced (or generated), transmitted, and used almost instantaneously. When any light switch is flipped or any machine is turned on, there is only a moment of delay between the creation of the needed electricity and its delivery to the waiting device. The ability to generate electricity on an as-needed basis in real time to satisfy demand for all homes, businesses, factories, and other users simultaneously is a massive challenge. The demand for electricity that the system must handle, or the *load*, must be anticipated in advance and then continually monitored and adjusted. This task is further complicated by the fact that, although the grid infrastructure itself is interconnected by more than 450,000 miles of high-voltage transmission lines,[31] the ownership and operation of the grid is fragmented. A complex web of more than 20,000 electric generating plants and over 9,000 publicly owned utility companies, private companies, and other entities are involved.[32]

To simplify the logistics of coordinating the efforts of so many disparate groups, voluntary regional control areas were established within the grid system. Generating plants and utilities in each area are monitored by oversight entities called Independent System Operators (ISO) or Regional Transmission Organizations (RTO).[33] These groups assume responsibility for "balancing" or monitoring and regulating the flow of electricity across their portions of the grid and also coordinating with each other to facilitate the flow of electricity between regions as necessary. Their vigil is constant: 24-hours per day, every day of the year.

The challenge of regulating eclectic load across the grid is further complicated by the age of the underlying infrastructure. Parts of the U.S. grid were first built in the 1890s when the electricity usage per home was far less than it is today because demand from most homes was tied to light bulbs and maybe a radio or television.[34] Today's modern device-assisted world includes much higher demands, albeit using increasingly more energy-efficient devices. The U.S. Energy Information Administration projects a continuing increase in demand for electricity between 2016 and 2040 across all sectors, including residential, commercial, and industrial users.[35] In that time, America's reliance on coal as an energy source is expected to decrease with a corresponding increase anticipated in the use of natural gas and renewable sources.[35]

This increase in electricity usage, regardless of the primary power source, creates challenges for the aging grid infrastructure, which already suffers from inefficiencies. Approximately 5 percent of all energy generated is lost in transmission between its point of origination and the end user.[36] In addition to distribution losses, often occurring through above-ground transmission lines that are not insulated, losses also result from power outages. These can happen for several reasons. For example, above-ground power lines are susceptible to storm damage and can cause extensive and costly power loss when damaged. System and monitoring failures can also cause significant power loss, such as occurred during the infamous 2003 Blackout that left 50 million people without power across the Northeastern United States and parts of Canada.[37] Additionally, the current grid system is also vulnerable to external cyber attacks, which have the potential to cause an even greater power loss event.

As part of the American Recovery and Reinvestment Act of 2009, an appropriation of $4.5 billion was made to update the grid.[38] In particular, the U.S. Department of Energy is working on plans to transform it into a "smart grid" that will use today's technology to actively monitor, regulate, and distribute energy flows rather than simply channeling electricity from source to customer.[39] Responsibility for the Grid Modernization Initiative is shared across several different Department of Energy divisions. There is a substantial cost involved in modernizing the grid, more than $2 trillion by some estimates.[40]

Gas. Natural gas is collected from shale fields at more than 500,000 wells across the United States through an extractive process, either horizontal drilling or hydraulic fracturing, commonly called "fracking."[41] The majority of natural gas collected and refined in the United States (34 percent of total production) is used to generate electricity.[42] In fact, natural gas is the leading primary energy source for electricity generation in the United States, ahead of both coal and nuclear sources.[43] The next highest sources of demand for gas stem from industrial and residential uses, at 29 and 16 percent of total production, respectively.[44]

Not unlike the electric grid infrastructure, natural gas is distributed by its own extensive infrastructure system, in this case a pipeline network, as shown in Fig. 2.5. Over 300,000 miles of wide-diameter, high-pressure interstate transmission pipelines and millions of miles of medium- and small-diameter distribution pipelines make up this system.[45] There is a specific naming convention for pipes in the network based on their size

FIGURE 2.5   U.S. electric and gas distributions lines. (*Source: U.S. Department of Energy.*)

and function. For example, *Hinshaw pipelines* are large pipes that receive gas from interstate suppliers but do not themselves operate outside of a particular state's borders.[46] Moderately sized pipes distributing through local service areas are referred to as *mains*; *service lines* are small pipes that connect directly to end user locations such as homes or businesses. On the whole, the pipeline network is newer than the electric grid. Most pipes were installed in the 1950s and 1960s, with further expansion in the 1990s and 2000s.[47]

Both electricity and natural gas are regulated across federal, state, and local levels—often by the same entities. At the federal level, the U.S. Department of Energy and the Federal Energy Regulatory Commission are both involved in regulating elements of the national electric grid as well as the production and sale of natural gas. The Federal Energy Regulatory Commission (FERC) is an independent government agency that oversees, among other things, the transmission and sale of electricity and natural gas across state lines.[48] Note that the commission's purview is limited to interstate activities; it does not regulate state and municipal power utilities or consumer sales of electricity and natural gas. FERC is also involved in the approval process for modifications to existing gas pipelines and construction of new gas pipelines. Electric utilities also work closely with the North American Electric Reliability Corporation (NERC), and the Environmental Protection Agency is often involved in natural gas development.

Both electricity and natural gas may be provided to consumers through a combination of wholesalers and local utility companies. The activity of these utility providers is regulated at the state level.

Implications for Development. As with the other infrastructure utilities already discussed, developers need to connect their projects to the national electric grid and sometimes to the natural gas pipeline network. The latter is especially true for the development of new industrial facilities with a heavy natural gas dependency, but may also be related to cooking and/or heating needs for residential projects. Factors that contribute to the developer's connection costs include (1) the

electric and/or gas draw the completed and occupied project requires to operate and (2) the proximity of the utility infrastructure.

The developer's need for these utilities will depend on the scope and size of the proposed project. In rural developments, infrastructure proximity is a concern for developments where the closest power lines may not be adjacent to the site. The cost per foot to run electric lines is different when the lines are run above ground than when they are below ground. Projects with large electric (and/or natural gas) demands may require the developer to work with the local utility provider to assess the capacity of the existing infrastructure. For very large projects, capacity upgrades may be required or the developer may have to include an on-site power substation as part of the development program. This cost may be attributed to the developer or may be shared with the utility company, depending on circumstances. For example, if the utility company believes the projected usage by the development project and resulting fees paid to the utility company will offset the long-term cost of the capital improvements, then the company will undertake the infrastructure work and significant upfront costs may not have to be borne by the developer. If, however, the utility company calculates that the additional, future fees paid by the developer will not adequately offset the cost to increase capacity, then the developer may have to pay upfront costs related to the infrastructure development. This can be a complex process requiring the developer to perform significant research, discussion, and negotiation during the due diligence period.

Social Infrastructure. The infrastructure categories discussed in the previous sections have all been examples of "hard" or "physical" infrastructure. This section will briefly address *soft infrastructure*, which includes those institutions that support society, such as schools, the police force, and fire departments. These are sometimes also referred to as *social infrastructure*. Unlike physical infrastructure, the need for social infrastructure can be difficult to quantify and, therefore, hard to plan for and accommodate. However, as with physical infrastructure, new development increases use of social infrastructure institutions and triggers the need to increase capacity.

While there is some debate about what exactly should be included in social infrastructure, schools are a generally accepted category. Schools also represent an area directly impacted by development, particularly new residential development. According to the U.S. Department of Education, 50.7 million children are enrolled in school between kindergarten and 12th grade.[49] That number is expected to increase steadily through 2023.[50] However, the location of the growth varies by state, as shown in Fig. 2.6, which can make it difficult for school systems to predict and fund increased staffing and capacity.

Implications for Development and Adequate Public Facilities Ordinances.   Although classified as a form of social infrastructure, school systems have significant physical components: school buildings, related recreation facilities, and the land they occupy. When suburban homebuilding increases the need for new or expanded schools, land is not always available on which new buildings can be built unless the jurisdiction has responsible growth plans in-place that anticipated these future needs. In some instances, developers may be asked to incorporate land for a new school into their site plans. However, adding new residential units in an urban environment is far more problematic as land to accommodate increased school capacity may not be readily available or affordable.

Many jurisdictions apply Adequate Public Facilities (APF) ordinances to development approvals in order to offset the impact of new development on infrastructure systems. APF ordinances are created by special legislative authority, particularly in communities where rapid growth has outstripped the government's ability to fund and/or otherwise provide necessary physical or social infrastructure. Typical APF ordinances require developers to delay or even forgo a project if the projected demand it would create for public services exceeds the total capacity of available funding for any of the systems

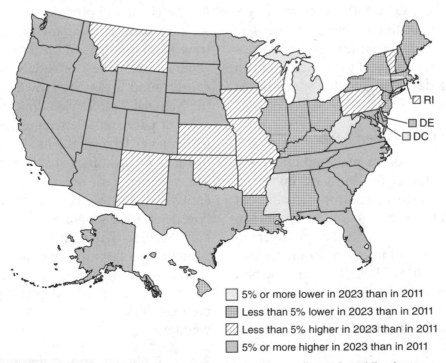

FIGURE 2.6   Projected percentage change in enrollment in public elementary and secondary schools, by state: fall 2011 through fall 2023. (*Source: U.S. Department of Education.*)

under review. In some instances, developers might be given the option to fund a shortfall or make improvements themselves, although this obviously increases the cost of the project and is not necessarily desirable from the developer's perspective. As an example of scale, the required school capital facilities charges in some suburban northern Virginia jurisdictions can exceed $50,000 per single-family home built. The infrastructure typically covered by APF ordinances includes transportation, water, sewer, and school systems. However, it can also be extended to include fire and police services because the availability and response times of these types of social infrastructure are tied to public health and safety.

APFs have certain drawbacks in practice. For example, in order to apply an APF ordinance test, the local government must create both a measure for the projected increase in demand as well as a determination about how to quantify adequate service. For "hard" services, such as sanitary sewer capacity, this is relatively easy to do using direct measurement and design capacity. However, for services such as school capacity or police support, it is dependent on a range of factors and can be far more difficult to accurately define. Further, to successfully use APF ordinances, local governments must also have accurate information about the existing capacity of different infrastructure systems. This can be difficult to track as current measures of capacity can be subject to change based on approved pipeline development that may not yet have been completed.

2.2.2. Parks and Open/Green Spaces

Parks and open/green spaces make up one of the three main categories of the built environment. They are not only critical environmentally, but they also contribute to the quality of the human experience. This can be difficult for developers to appreciate because development generally focuses on the best *economic* use of land; however, the best societal or environmental use of land is an entirely different, yet important, measure.

Although parks and open/green spaces are largely undeveloped, they are considered part of the built environment because deliberate decisions were made *not* to build on them. For example, zoning laws, which are discussed more fully in the next section of this chapter, often have specific landscaping and open space requirements. Note that not all open spaces are necessarily "green" spaces. The distinction is dependent on the percentage of natural coverage included in the open space. A paved urban plaza is an example of an open space that is not necessarily green, while a grassy neighborhood park is an example of an open space that is also green.

The open space requirements in zoning regulations can provide gathering places for a community or, when a green component is required, can also ensure the inclusion of nature in increasingly urban environments. In addition to providing ambiance, open green space requirements are also an important tool for managing the total impervious surface of the built environment. When surface area is completely covered with impervious structures such as buildings, roads, and parking lots, it prevents rain water from being reabsorbed by the earth, which impacts the natural water cycle and increases stormwater management problems.

Parks are also important to consider. The 1916 "Organic Act" created the National Park Service to oversee the fledgling chain of American national parks and provide for the conservation of "the scenery and the natural and historic objects and the wild life therein and to provide for the enjoyment of the same in such manner and by such means as will leave them unimpaired for the enjoyment of future generations."[51] The size of U.S. national parks and conservation areas has grown since 1916. Today, the National Park Service alone oversees 84 million acres of parkland. That represents an area almost *600 times* as large as the total square footage of all privately owned office space in the Unites States.[52] The National Park Service is not the only federal entity that controls land. Another example is the Forest Service, a separate agency from the National Park Service that manages its own portfolio of 54 national forests and 20 grassland areas.[53]

State and municipal governments also own and maintain parks and playing fields, although on a significantly smaller scale than their federal counterparts.

2.2.3. Buildings and Structures

Buildings and other structures represent the most obviously recognizable component of the built environment. As a category, however, "buildings" is too large to be particularly useful. For example, both a single-family home with a large yard and a 25-story high-rise office in an urban environment are both "buildings," yet they have very little in common with respect to design, cost, or function. Only the largest developers have the necessary competence to undertake projects of both types. As a result, buildings are often subclassified based on their purpose, scale, or height. As introduced in Chap. 1, the typical use categories, or *product types*, are residential (single or multifamily), retail, office, and industrial. When categorizing buildings by height, generally those between a single story and four stories in height are referred to as *low-rise* (or *garden style* for apartments); buildings between five and 10 stories are *mid-rise*; and buildings with more than 10 floors are considered *high-rise* buildings.

Buildings can also be classified by quality based on their age, condition, and level of amenities. A poorly maintained building that is 50 years old would be loosely classified as *Class C*, while a brand new, modern building incorporating the latest sustainable and smart building technology would be considered *Class A*. Class C buildings represent an interesting component of the built environment. On the one hand, by virtue of their age and lack of amenities, they tend to command lower rental rates and may be more affordable than other options. As such, have an important social and economic role in society. On the other hand, they represent underutilized value for their owners and are the primary targets of certain specialty developers, who seek out well-located Class C buildings in order to redevelop them. Note that while any building can be referred to by class, the practice is most commonly used with respect to office buildings.

Different members of the development team view buildings through their respective professional lenses. For example, the civil engineer might think of a building in terms of its structural integrity, life systems, and layout. An attorney might only be concerned with lease clauses and areas of liability for the owner. Developers may look at buildings as sources of economic opportunity, but often consider them in the broadest sense.

Public Sector Building Portfolio. Though it is often overlooked, the government owns a vast and diverse portfolio of buildings. At both the state and municipal level, this typically includes office buildings; schools and training facilities; and police stations, courthouses, and jails. However, state and local governments can also own retail properties, medical facilities, laboratories, warehouses, and other types of buildings.

The federal portfolio of buildings is substantial. All executive branch agencies are required to make annual reports of their real estate portfolios as part of the Federal Real Property Profile (FRPP) Management System.[54] According to the 2017 FRPP, the executive branch alone owned 238,003 buildings comprising a total of 2.399 billion square feet.[55] These measures do not include leased space, or any buildings owned or leased by the Department of State overseas. With some exceptions, they also exclude non-executive branch portfolios, such as the United States Postal Service, Smithsonian museums, and legislative and judicial branch buildings.

The public sector portfolio is important to note because, as with the infrastructure projects previously described, all of the government's buildings were development projects at one time. As agencies at each different level of government grow, shrink, consolidate, or relocate, their building and space requirements constantly change. This creates yet another public sector need for development on a large scale.

2.3. PLANNING FOR THE BUILT ENVIRONMENT

As the previous section demonstrated, the built environment is complex, comprised of critical infrastructure, a vast range of building types, open spaces, and other features. Although historically there was little regulation to guide the development of these different uses of land, today they do not occur haphazardly. The coordination of how the different components of the built environment come together falls under a specialized discipline called *planning*. Professional planners seek to address the current and future needs of society and translate solutions into plans for land use. Planning exercises can focus on different geographies, such as regional or urban planning, or different specialty areas, such as transportation planning.

Best practices suggest that planning efforts should be coordinated across different jurisdictions, particularly with respect to regional transportation planning, although this does not always happen in practice.

This section will discuss the need for planning, the planning process, and the use of zoning to produce planning results.

2.3.1. Regional and Transportation Planning

Regional planning is a voluntary, collaborative exercise that crosses jurisdictional boundaries. Depending on the size of the area in question, regional planning efforts can involve different municipalities, counties, or even states. At the local level, city and county governments, elected officials, transportation departments, planning commissions, and transit agencies are included. State level participants can include governors, state legislators, and state departments of transportation. Federal entities are also usually involved when projects require federal funding, including the U.S. Department of Transportation (Federal Highway Administration, Federal Transit Administration), the U.S. Environmental Protection Agency, the Department of the Interior, and even Congress.

Sound regional planning is necessary to accommodate the needs of growing populations as well as to account for aging highway and transit infrastructure systems that often suffer from an extensive backlog of deferred maintenance and limited funding resources. Regional planning is particularly important in the context of the twenty-first century's globally competitive economy, where regions that build and sustain well-connected, multimodal transportation networks are more likely to be economically successful than those that do not. Essential components of a region's success include thorough planning; active consensus building; a well-defined implementation strategy; and the allocation of local, regional, state and federal funds to projects that best promote the regional movement of people, goods, and services. Historically, regional planning exercises have focused on maintaining infrastructure continuity and capacity rather than designating specific land uses; thus, this section will be limited to a discussion of regional planning with respect to transportation planning.

Metropolitan Planning Organizations. Historically, transportation planning was problematic for local and state governments, which often struggled in isolation or may have found themselves at odds with the competing goals, needs, wants, and interests of neighboring jurisdictions. However, the 1962 Federal Highway Act mandated the establishment of Metropolitan Planning Organizations (MPOs) for urban areas with populations greater than 50,000. These organizations serve to encourage consensus building on transportation planning and the allocation of federal funds for specific projects, thus helping coordinate efforts between disparate jurisdictions to facilitate planning progress.

MPOs are comprised of representatives appointed by municipalities, counties, regional transit authorities, and other transportation agencies within the urban area as well as any relevant state transportation agencies. They include both elected officials and government employees. The appointed members are responsible not only for planning initiatives, but also for the governance of the MPO entity. Members of one or more federal transportation agency are also usually involved in MPO activities in a nonvoting capacity. Note that in some regions where there are also subregional transportation planning organizations, best practices dictate that such entities also be included in MPO activity in some capacity.

Most MPO's have several advisory committees, usually including a Technical Advisory Committee (TAC) as well as a Citizens Advisory Committee. The TAC consists of representations from local and state transportation departments appointed by the MPO governing body as well as MPO professional staff. The main responsibility of the technical committee members is evaluating and making recommendations regarding project, planning, and policy proposals. Citizen committees facilitate outreach activities and provide feedback related to MPO policies.

Typical MPO responsibilities include both long- and short-range planning; forecasting travel demand as well as housing, employment, and population trends; and monitoring air quality conformity. Long-range planning involves the formulation of a fiscally Constrained Long Range Plan (CLRP). The element of "fiscal constraint" was imposed by the 1991 Intermodal Surface Transportation Efficiency Act (ISTEA) to promote realism and discourage populating long-range plans with scores of projects having unrealistic funding requirements. CLRPs generally cover a 25-year period and consist of recommendations for highway, transit, and other projects for which funding can be reasonably identifiable. MPOs also arrange for regular testing to ensure that projects in the CLRP collectively contribute to federal air quality improvement goals. Forecasting travel demand, population, employment, and housing trends is a crucial component to successful long-term regional planning and MPOs continually work on the development of data for use in regional transportation and modeling exercises. Projections are updated regularly and usually forecast trends over a period of 20 to 25 years.

An MPO's short-term planning activities focus on the creation of Transportation Improvement Programs (TIP), which cover a period of at least four years. TIPs are prepared in cooperation with local and state transportation agencies, who may have individual members within the MPO but which are not organizational members in their own right. TIPs must be consistent with the CLRP; fiscally constrained; include highway, transit, bicycle, and pedestrian components and safety investments; and account for all regionally significant projects receiving federal highway and transit funding.

The projects evaluated and recommended by the MPO are selected based on nominations from projects included in local, state, and area transit agency plans. The requirement that funding must be reasonably available is, of course, a fundamental consideration. New projects are generally submitted for consideration in conjunction with regular CLRP and TIP updates, although exceptions can occur when unexpected funding becomes available or other foreseen events arise. In many cases a given project's merits are well documented; however, geographical, modal, ideological, or other factors can be powerful counter balances. For example, a project with demonstrable congestion

or travel time reduction benefits may not necessarily be selected if the majority of MPO members prioritize other concerns such as bicycle and pedestrian projects designed to discourage automobile usage. MPOs operate on the basis of unanimity and, therefore, seek 100 percent consensus in order to move forward with recommendations. In many instances they succeed; however, opposition from a minority of members with special interests may result in the rejection of a project of true regional significance due to lack of complete agreement.

Unsurprisingly, some of the main challenges facing MPOs revolve around ensuring an impartial process that is uninhibited by political maneuvering. While the MPO entity is charged with regional planning and inclusion of projects of regional significance, the actual MPO members, who are appointed at the local level, ultimately make decisions about project selection. In their decision-making, members inherently face conflict between the regional obligations of the MPO and supporting their local communities. Members that are locally elected officials (as opposed to permanent government employees) may be further swayed by reelection pressures to designate funds to projects considered locally important rather than to projects beyond their own municipal or jurisdictional borders. This tension between priorities can call into question whether MPO recommendations truly address regional needs or are merely a collection of negotiated trades for locally preferred projects that, in fact, may not be particularly relevant in the context of a true regional plan. Similar struggles manifest themselves across a range of MPO considerations and raise questions such as:

- *Modal "balance."* Are funds awarded to improvements for specific transit modes (auto/transit/freight/pedestrian/bicycle) that move the most people regardless of mode or are they distributed across modes regardless of level of usage?

- *Assessment of "fair share."* Are projects selected and funds distributed based on greatest need or distributed equally among jurisdictions regardless of need?

- *Criteria weighting.* Is the objective to promote regional mobility by moving the greatest number of people or to advance other matters? (While congestion reduction and improving fundamental connectivity were once primary considerations, today a multitude of measures are considered including land use, safety, environmental, and mode choice. The amount of weight or importance assigned to such measures can greatly affect/alter project selection.)

Accountability is another challenge for MPOs. While MPOs and their members are charged with selecting and allocating funds to specific projects, they may or may not be held accountable for how well these projects perform. Such accountability generally falls to municipal or state transportation departments, which may or may not have agreed with the selection of the projects they are tasked to carry out. Election cycles can also contribute to accountability issues. Facts and well-documented regional needs seldom change, but MPO members do. Local and state elections can change jurisdictional transportation agendas, sometimes significantly if elected officials inject personal views into MPO decision making through their support for the selection of different projects. Ideological differences have become greater factors as changes in federal law and other forces have effectively broadened the MPO's role to include land use, environmental protection, and auto-alternative mode promotion. This can lead to decision making that results in project selection based on how some MPOs members think the public should live and travel as opposed to how the public actually lives and travels.

In the worst case, if an MPO is unable to overcome prioritization challenges, it can result in the development of a long-term regional plan and selection of projects that reflect local goals as opposed to true regional needs, does not suitably address congestion and travel times, and favors special interest views over the judgment of transportation professionals. Such a failing is clearly a bad result. However, many MPOs are successful in using the best available objective data to produce meaningful results through the development of a long-term, regionally focused plan and the selection of projects that result in improved regional mobility. It should be noted that the exclusion of a locally important project from the regional plan does not mean the project is wholly unimportant or insignificant; rather, that the project does not rise to the level of regional significance and should be funded and constructed by the local jurisdiction.

2.3.2. Comprehensive Planning

The majority of planning exercises that impact developers happen at the local level where county or city governments have an immediate responsibility to citizens in their particular jurisdictions. The conditions, needs, and challenges facing different municipalities, even within the same state, can vary tremendously, leading each to focus on a unique set of problems and priorities. This may include things such as the need for job growth, revitalizing a dwindling industrial or retail corridor, the need for historic preservation, managing rapid expansion, or accounting for the impacts of changing technology and lifestyle preferences. The comprehensive planning process attempts to capture and support the economic, social, business, demographic, and environmental needs of a community. An important goal of the process is to allow both public and private sector stakeholders to contribute input about their needs and aspirations. This is often achieved through an inclusionary program that may include direct outreach, charrettes, open input opportunities, and public meetings. Of course, planners also consider demographic and employment data, infrastructure considerations, environmental issues, housing supply, and other data when building their recommendations.

Properly executed planning can enhance the relationship between various land uses and improve the lives of citizens. A successful planning process can also lead to increased economic efficiency by coordinating the size and location of physical features with projected future needs. Due to the complexity of the task and reliance on often-limited resources, large area planning exercises can take several years to complete.

When finished, the results are used to create a *comprehensive plan*, sometimes also called a *master plan* or *general plan*. A comprehensive plan captures a community's vision for the future and is a long-term guide for growth and development. Usually, the comprehensive plan will include dedicated sections, chapters, or subplans on important thematic areas such as housing, the environment, transportation, community design, and economic development. Each of these elements will describe priorities and policy recommendations relevant to their particular area of focus. Collectively, the comprehensive plan establishes policies and procedures relating to a community's future growth, including new development of land and maturation of existing built areas, usually for a period of 20 years.

Typical components of a comprehensive plan include:

1. Statements about the community's goals and objectives.

2. Inventories of its existing characteristics, features and resources, land uses, and facilities.

3. Projections of trends expected within the life of the plan.

4. Text describing policies to be applied in order to achieve the plan's goals.

5. Maps and text depicting and discussing the community, showing current and future land use, the location of future public facilities, environmental resources, and other features.

6. Implementation text describing how the community intends to carry out the goals of the plan.

Good comprehensive plans include focused, specific goals that can be translated into actionable results. These might include policy recommendations to incentivize growth in specific parts of the community or the building of certain uses by offering developers density bonuses, tax credits, or special low-interest loans.

Accompanying the comprehensive plan text is a proposed land use map, such as the one shown in Fig. 2.7. Often color-coded or shaded, the generalized land use maps of the community represent a graphic depiction of the relationship of existing and future land uses and facilities. The map divides geographic areas into desired and projected uses and intensity. These areas usually represent the broad categories of land use, such as residential, commercial, industrial, and other employment centers. Subcategories of development intensity show the gradation of land use patterns. For instance, the map may show that a high-density residential area is a desirable future use adjacent to a commercial center. The map defines the boundaries of the area and provides a range of relative densities. It will also frequently show the proposed locations of significant facilities, regional shopping malls, and schools.

Old comprehensive plans can be updated, modified, or even redone if necessary. Usually, a comprehensive plan, or some of its components, will be updated every five to ten years to ensure that changes in community needs are properly accommodated. Comprehensive plans can also be supplemented with targeting planning exercises focused on smaller, specific geographic areas. These may include business districts, industrial areas, transit centers, or other significant areas and are called *sector plans* or *small area plans*. Comprehensive plans are sometimes also supplemented with individual functional plans that focus on important areas such as transportation, water and sanitary sewer, or environmental issues.

It is important to note that comprehensive plans are simply policy recommendations to guide a community's future growth and not a firm declaration of the future. In most states, comprehensive plans serve only as guides and *not* as legally binding resolutions that require strict conformance. It is also worth noting that the guidance provided by a comprehensive plan is only useful within the boundaries of the particular municipal jurisdiction to which the plan belongs; an adjacent jurisdiction may have an entirely different set of needs that result in the creation of a materially different comprehensive plan. Ultimately, these plans only impact development projects if either (1) elements of the comprehensive plan are incorporated into zoning laws, or (2) if the developer seeks a change of zoning in a jurisdiction that intentionally incorporates strict adherence to the comprehensive plan into its approval process. The next section will discuss zoning in more detail.

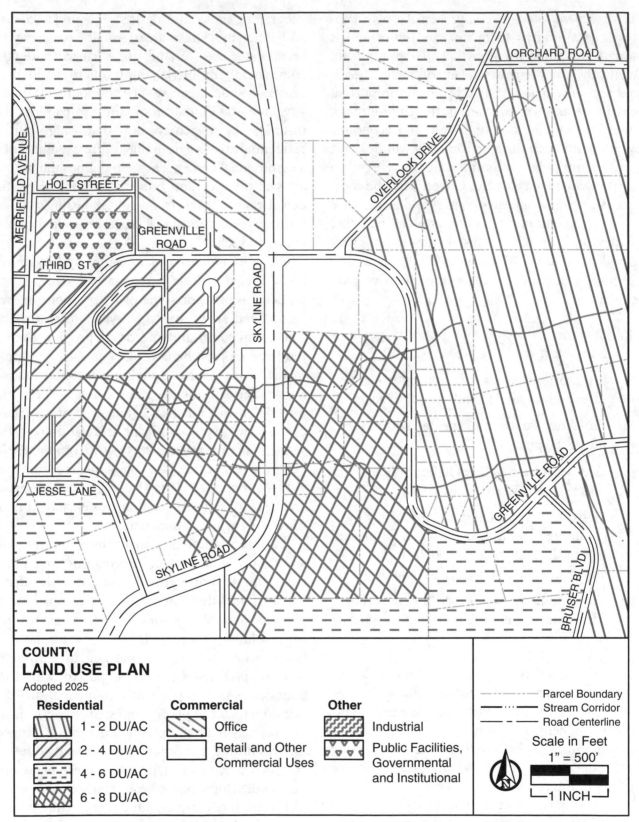

FIGURE 2.7   Sample land use map.

2.3.3. Zoning

Planning is an important activity for expressing the future vision of a community. As an isolated exercise, however, it has little actual influence on the creation of the built environment. As mentioned in the previous section, comprehensive plans in most states are only guides for development. Local governments require a mechanism to harness the work and wisdom captured in a comprehensive plan and translate it into reality. This is made possible by a special kind of governmental authority, called *police power*, which allows governments to pass (and enforce) laws and regulations in order to safeguard public health, safety, and welfare. In this case, use of the word "police" obviously refers to the administrative act of "policing" rather than to actual law enforcement officers. This power allows the government to make decisions that prevent overcrowding, establish appropriate sanitary regulations, provide for an efficient transportation system, and protect quiet residential areas.

The regulation of land use under police power authority is achieved by passing zoning laws, also called *ordinances*, which restrict and control how land can be used. A zoning ordinance is the legislative means by which a municipal jurisdiction sets detailed requirements for all aspects of land use. For any parcel of land, currently developed or otherwise, these ordinances can specify type of allowable use, layout, density, height, setbacks, parking requirements, open space requirements, signage standards, and impervious area requirements. Zoning laws are instituted by the authority of a local legislative body, such as a county board, city, or town council and recorded as part of the jurisdiction's code of laws. It is important to note that zoning ordinances are not required to follow the recommendations of the comprehensive plan.

Zoning, as the name suggests, is a process by which a geographical area is divided into separate zones, each of which is designated for a particular kind of use. This allows zoning ordinances to prevent conflict by creating separation between incompatible uses, such as a school and a liquor store. Sometimes the desired separation can be created by physical distance, but this is not always possible, especially in dense urban environments or at the boundaries between different zones. In such cases, zoning requirements may attempt to achieve the same goal through architectural solutions such as smaller signage, larger yards or setbacks, fencing, and landscaping requirements.

Historically, there were only three major zoning categories: residential, commercial, and manufacturing/industrial. Each of these property uses were isolated and allocated to distinct areas of a municipality through a rigid single-use zoning regime known as *Euclidean zoning*. Under Euclidean zoning concepts, different property uses are not commingled. Although much of the country still uses this form, zoning has evolved considerably and today there are a wide range of zoning strategies and use-categories. These contribute to the vibrancy of the modern built environment by accommodating different circumstances, allowing different interaction between uses, or even interweaving different compatible uses together. Some of the newer, more flexible zoning concepts include form-based zoning and mixed-use zoning.

From a developer's perspective, while some zoning regimes may be easier to work within than others, the type of zoning regime itself is largely irrelevant. Developers must inherently work within whatever type of zoning scheme a municipality puts in place. A developer will not abandon a well-located site that represents a viable development opportunity simply because it is subject to Euclidean zoning as opposed to mixed-use zoning. Equally, a developer that is familiar with a particular market is unlikely to abandon pursuit of all future deals in his/her home market in favor of an unfamiliar market with a more flexible zoning regime.

While the overarching type of zoning regime may not impact a developer's choices, the specific zoning *designation* of a particular site most certainly will. The zoning ordinance itself will specify what uses are permitted on a particular site because they comply with the intent of the zone or zoning district. These are referred to as *by right* uses. In addition to by right uses, the ordinance language will also list uses that may be allowed if certain conditions apply (if any); uses that will be permitted by special exception (if any); uses permitted with a special-use permit (if any); and, in some cases, specifically prohibited uses. *Special exceptions* can be granted in cases where the proposed use may be generally consistent with the intent of the zoning district but have potential impact on surrounding properties or the community that requires special consideration. A common example is building a church or other place or worship in a residential zone. Approval for a special exception may be tied to additional restrictions intended to help mitigate any impact on the community. It is possible to deviate from zoning ordinances in certain situations through a request for a *variance*. Although use-variances exist in certain jurisdictions, the most common form of variance is tied to the physical limitations of a site that make it difficult or impossible to meet existing zoning requirements, such as setback distances. Note that a developer's chances of success for approval depend on the specifics of the variance being requested. In some cases, if zoning prohibits the developer from building a particular project, he/she may choose to seek a formal *rezoning* in order to accommodate the desired use. This can be a long, risky process that is not always worth pursuing for all projects.

Most zoned areas are still dedicated to single uses, especially in nonurban areas. These zones can be broken down into smaller zoning districts, which accommodate variations on the types of intensities of similarly zoned uses. For example, a residential zone, usually (but not always) denoted with "R," might include specific zoning districts for low-density development, medium-density development, and high-density development. *Density* refers to the number of dwelling units or square feet of space built on a set area, usually defined in acres. The zoning for each district will specify what types of buildings are permitted, the minimum lot size, and number of dwelling units permitted per acre—all of which contribute to a site's density. For instance, a site that permits only single-family detached homes on 1-acre lots will have a much lower density than an urban site that allows a 15-story multifamily building with 150 apartment units. The zoning regulations may also specify a maximum lot size, building height, distance between buildings, and setback from lot lines and roadways.

These bulk zoning restrictions are especially important to a developer and his/her technical team members because they dictate how much building space can be accommodated on each site. For nonresidential projects, this measurement is

known as the *floor area ratio* (FAR). FAR is the ratio of total building square feet (sf) to total site area. Note that FAR is independent of a building's height or number of stories. For example, a 40,000 sf building constructed on a 40,000 sf lot has an FAR of 1.0. The building may be comprised of 20,000 sf on two floors or 10,000 sf on four floors, but the FAR does not change. FAR is critically important to developers because the total size of the building, more specifically its useable space, dictates how much square footage the developer has available to rent or sell. Developers will almost always want to maximize FAR. To the degree that a municipality seeks to stimulate growth and new development, it can attract developers through its zoning ordinances by allowing higher density development projects than neighboring jurisdictions.

Some zoning ordinances accommodate a mechanism that allows developers to transfer density between unrelated sites. This is achieved through a specially legislated commodity known as a *transferable development right* (TDR). Through the use of TDRs, the owner of one property can sell unused density to the owner of another property, who can then develop a larger project than he/she might have been able to by right. TDRs serve as a planning tool that can help preserve lower density development in protected parts of a community, such as historic areas or to preserve green/open areas, without penalizing owners economically for development restrictions imposed in those areas. Simultaneously, TDRs can incentivize higher density development in targeted growth areas. These are often known, respectively, as "sending" and "receiving" areas within the comprehensive plan.

To illustrate zoning constraints, a sample R-12 medium-density residential zoning regulation is provided in Fig. 2.8, and a hypothetical 2-acre undeveloped site is provided in Fig. 2.9. The site is located in a suburban area on the south side of Atherton Road and will be referred to as "Site A" throughout the rest of this book. Site A is known as an *infill* site because it is vacant or underutilized land already surrounded by existing, developed lots. Using this information, we can briefly track the basic options facing a developer interested in building on the site. By right, the developer can build as many as 24 dwelling units, assuming the open space and parking requirements are also met. If the developer wanted to include a child care center as part of the development, he/she would need to seek a special exception. If the developer wanted to build something else on the site, such as a retail center, he/she would be required to seek a change of zoning for the site.

The zoning of a site impacts the type, density, and style of a future building. This has obvious ramifications for developers. As previously explained, zoning is a form of land use control. By extension, it may sometimes be viewed by the developer as also "controlling" the developer's opportunities, projects, and profits. From the developer's perspective, the municipal entities responsible for passing

zoning laws often employ a broad interpretation of their role in promoting public health and welfare and, thus, can impose restrictions that are more stringent than necessary. For example, zoning restrictions that focus on aesthetic standards are, arguably, not strictly necessary to promote health, safety, and welfare. Regardless, the developer will not receive project approvals or permits to build if he/she does not satisfy zoning requirements.

Zoning often does not apply to public sector developers in exactly the same way that it does for private sector developers. For example, most zoning designations usually include a broad special allowance for "public uses," as shown in Fig. 2.8,

---

R-12 RESIDENTIAL DISTRICT, TWELVE DWELLING UNITS/ACRE

Purpose

The R-12 District is established to provide for a planned mixture of residential dwelling types at a density not to exceed twelve (12) dwelling units per acre; to allow other selected uses which are compatible with the residential character of the district; and otherwise to implement the stated purpose and intent of this Ordinance.

Permitted Uses

    1. Dwellings, single family attached
    2. Dwellings, multiple family
    3. Dwellings, mixture of those types set forth above
    4. Churches, chapels, temples, synagogues, and other such places of worship
    5. Public uses

Special Permit Uses

    1. Convents, monasteries, seminaries, and nunneries
    2. Subdivision and apartment sales and rental offices
    3. Temporary dwellings or mobile homes
    4. Temporary farmers' markets

Special Exception Uses

    1. Child care centers and nurseries
    2. Cultural centers and museums
    3. Independent living facilities
    4. Funeral homes

Lot Size Requirements

    1. Minimum district size: 2 acres
    2. Minimum lot area:
        A. Non-residential uses: 10,000 sf

Maximum Density

12 dwelling units per acre

Open Space

25% of the gross area shall be open space

Parking Requirements

2 dedicated spaces per dwelling unit with 1 additional visitor space per dwelling unit

---

FIGURE 2.8 Sample residential district regulations. (*Source: Based partially on Fairfax County, VA codes.*)

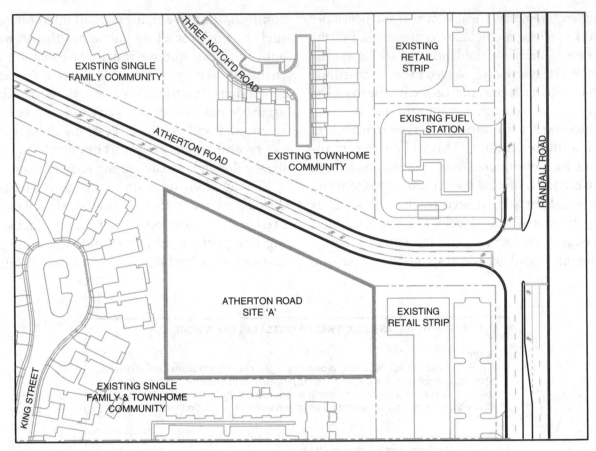

FIGURE 2.9   Atherton Road site.

which gives the public sector developer far more leeway in identifying project sites. Additionally, the government has the right to simply take land and use it for public sector projects. The ability to expropriate private real estate (or other private property) for public use is a special form of police power called *eminent domain*. The act of the taking itself is referred to as a *condemnation*. While the government *is* required to pay fair and just compensation to the owner for the land or property taken, the taking itself is compulsory for legitimate public purposes. Citizens have challenged the legality of certain takings on the basis that the government did not demonstrate a sufficient public cause to justify the action, but eminent domain powers have generally been supported by the courts. The power of eminent domain is imbued across federal, state, and local governmental levels, but is perhaps most commonly seen in state-level takings for road infrastructure projects.

CHAPTER CONCLUSION

This chapter provided detail about the complex built environment in which developers conduct their business. With this important context now established, it is appropriate to examine the stages of the development process in greater detail. The next chapter will discuss key considerations related to Milestone 1 (recognizing opportunity) and describe different project types that a developer may consider. It will also describe the considerations of different development team members for different project types.

REFERENCES

1. ASCE 2009 Infrastructure Report Card website available at https://www.infrastructurereportcard.org/2009/report-cards.html (accessed April 26, 2018); ASCE 2013 Infrastructure Report Card website available at http://2013.infrastructurereportcard.org/ (accessed April 26, 2018); ASCE 2017 Infrastructure Report Card website available at https://www.infrastructurereportcard.org/ (accessed April 26, 2018).

2. ASCE 2017 Infrastructure Report Card website, section on Economic Impact available at https://www.infra-structurereportcard.org/the-impact/economic-impact/ (accessed April 26, 2018).

3. FY 2016 Municipal, Industrial and Irrigation Water Supply Database Report 2017-R-02, The Institute for Water Resources (IWR), U.S. Army Corps of Engineers (USACE), June 2017, available for download at https://publibrary.planusace.us/#/document/e71cd9b9-faff-4d63-8d42-8ee1baa3cc7f (accessed October 14, 2018); City of New York, Department of Environmental Protection Open Data set on water consumption, available at https://data.cityofnewyork.us/Environment/Water-Consumption-In-The-New-York-City/ia2d-e54m (accessed September 8, 2017).

4. FY 2016 Municipal, Industrial and Irrigation Water Supply Database Report 2017-R-02, The Institute for Water Resources (IWR), U.S. Army Corps of Engineers (USACE), June 2017, available for download at https://publibrary.planusace.us/#/document/e71cd9b9-faff-4d63-8d42-8ee1baa3cc7f (accessed October 14, 2018).

5. U.S. Army Corps of Engineers, National Dam Inventory open data set interactive report, available at http://nid.usace.army.mil/cm_apex/f?p=838:4:0::NO (accessed September 8, 2017).

6. Washington Aqueduct, District of Columbia, VA, and MD, Fact Sheet as of March 14, 2018, U.S. Army Corps of Engineers; Arlington Virginia, Water and Utilities website, available at water.arlingtonva.us/water/ (accessed October 14, 2018).

7. Washington Aqueduct website, available at nab.usacc.army.mil/Missions/Washington-Aqueduct/ (accessed Aug. 18, 2017); Arlington Virginia, Water and Utilities website, available at water.arlingtonva.us/water/ (accessed August 18, 2017).

8. Arlington Virginia, Water and Utilities website, available at water.arlingtonva.us/water/ (accessed August 18, 2017).

9. EPA-832-B-00-002. Guidance on the Privatization of Federally Funded Wastewater Treatment Works; August 2000.

10. Water and Wastewater Systems Sector-Specific Plan, 2015, Department of Homeland Security and the U.S. Environmental Protection Agency, available for download at https://www.dhs.gov/sites/default/files/publications/nipp-ssp-water-2015-508.pdf (accessed October 2014, 2018); Department of Homeland Security website, Critical Infrastructure Sectors, Water and Wastewater Systems Sector, Last Published Date: July 6, 2017, available at https://www.dhs.gov/water-and-waste-water-systems-sector (accessed September 9, 2017).

11. Federal Register Vol. 57, No. 86, May 4, 1992, pp. 19063–19248; EPA-832-B-00-002. Guidance on the Privatization of Federally Funded Wastewater Treatment Works; August 2000.

12. Primer for Municipal Wastewater Treatment Systems, U.S. Environmental Protection Agency, Offices of Water and Wastewater Management, 2013, Washington, DC, available for download at https://www3.epa.gov/npdes/pubs/primer.pdf (accessed October 14, 2018).

13. EPA website on NPDES, available at https://www.epa.gov/npdes/about-npdes (accessed September 9, 2017).

14. EPA website on NPDES, Stormwater Discharges from Construction Activities, available at https://www.epa.gov/npdes/stormwater-discharges-construction-activities (accessed September 9, 2017).

15. Transportation and Infrastructure Committee website, available at https://transportation.house.gov/about/history.htm (accessed September 9, 2017).

16. U.S. Department of Transportation website, section "About Us" available at https://www.transportation.gov/mission/about-us (accessed August 19, 2017); Department of Transportation Combined FY 2015 Annual Performance Report and FY 2017 Annual Performance Plan, available for download at https:// www.transportation.gov/mission/budget/fy2017-annual-performance-plan-fy-2015-perf-report (accessed December 16, 2018).

17. Federal Highway Administration Office of Highway Policy Information, Highway Statistics report 2015 (most current data), available at https://www.fhwa.dot.gov/policyinformation/statistics/2015/ (accessed October 14, 2018).

18. The system's complete name is the "The Dwight D. Eisenhower National System of Interstate and Defense Highways" as per Public Law 101-427 of October 15, 1990; Federal Highway Administration website, MAP-21, Questions & Answers, section on National Highway System Questions & Answers, available at https://www.fhwa.dot.gov/map21/qandas/qanhs.cfm/ (accessed September 10, 2017); Federal Highway Administration website, Highway History, section on " Building the Interstate," available at https://www.fhwa.dot.gov/infrastructure/build02.cfm (accessed September 9, 2017).

19. Combined FY 2015 Performance Report and FY 2017 Performance Plan, DoT, p. 48.

20. Federal Highway Administration Office of Federal Lands Highway website, available at https://flh.fhwa.dot.gov/ (accessed September 10, 2017); Federal Highway Administration Office of Highway Policy Information, Highway Statistics report 2015 (most current data), available for download at https://www.fhwa.dot.gov/policyinformation/statistics/2015/ (accessed September 10, 2017); Federal Highway Administration website, section on Highway History, Interstate Frequently Asked Questions, available at https://www.fhwa.dot.gov/interstate/faq.cfm#question1 (accessed September 9, 2017).

21. DoT Federal Highway Administration publication "Creation of a Landmark: The Federal Aid Road Act of 1916," by Richard F. Weingroff available for download at https://www.fhwa.dot.gov/highwayhistory/landmark.pdf (accessed September 10, 2017).

22. Federal Highway Administration website, "The Trail-blazers: Brief History of the Direct Federal Highway Construction Program," available at https://www.fhwa.dot.gov/infrastructure/blazer01.cfm (accessed September 15, 2017).

23. Federal Highway Administration website, section "About:, available at https://www.fhwa.dot.gov/about/ (accessed September 9, 2017).

24. National Highway Performance Program (NHPP) Implementation Guidance, March 9, 2016, Federal Highway Administration, available online at https://www.fhwa.dot.gov/specialfunding/nhpp/160309.cfm#ProgramPurpose (accessed October 14, 2018).

25. 23 U.S.C. 120; National Highway Performance Program (NHPP) Implementation Guidance, March 9, 2016, Federal Highway Administration, available online at https://www.fhwa.dot.gov/specialfunding/nhpp/160309.cfm#ProgramPurpose (accessed October 14, 2018); "Funding Federal-aid Highways," Publication No. FHWA-PL-17-011, January 2017, Office of Policy and Governmental Affairs, Federal Highway Administration.

26. Federal-Aid Highway Act of 1956, Section 209.

27. Federal Highway Administration FAST Act Fact Sheet, available at https://www.fhwa.dot.gov/fastact/factsheets/htffs.cfm (accessed September 10, 2017).

28. Federal Highway Administration Office of Policy Development, Highway Trust Fund Primer, November 1998, available for download at https://www.fhwa.dot.gov/aap/PRIMER98.PDF (accessed October 14, 2018).

29. Congressional Budget Office Projections of Highway Trust Fund Accounts as of July 5, 2017, available at https://www.cbo.gov/sites/default/files/recurringdata/51300-2017-06-highwaytrustfund.pdf (accessed September 10, 2017).

30. Department of Transportation FY-2015 Annual Performance Report / FY-2017 Annual Performance Plan, available for download at https://www.transportation.gov/mission/budget/fy2017-annual-performance-plan-fy-2015-perf-report (accessed October 14, 2018).

31. Department of Energy website, "9 Things You Didn't Know about America's Power Grid," available at https://energy.gov/articles/top-9-things-you-didnt-know-about-americas-power-grid (accessed September 16, 2017).

32. U.S. Department of Energy, Office of Electricity Delivery and Energy Reliability website, Information Center, available at https://energy.gov/oe/information-center/educational-resources/electricity-101#sys1 (accessed September 16, 2017); U.S. Energy Information Administration 2015 (most recent available) data set on largest utility providers, available for download at https://www.eia.gov/energyexplained/index.cfm?page=electricity_home#tab2 (accessed October 14, 2018); U.S. Energy Information Administration website, Form EIA-860 detailed data, available at https://www.eia.gov/electricity/data/eia860/ (accessed October 14, 2018).

33. U.S. Department of Energy, Office of Electricity Delivery and Energy Reliability website, Information Center, available at https://energy.gov/oe/information-center/educational-resources/electricity-101#sys1 (accessed September 17, 2017); U.S. Energy Information Administration website, available at https://www.eia.gov/todayinenergy/detail.php?id=27152 (accessed September 17, 2017).

34. U.S. Department of Energy, Office of Electricity Delivery and Energy Reliability Smart Grid website, section on "What is the Smart Grid?" available at https://www.smartgrid.gov/the_smart_grid/smart_grid.html (accessed 26 April 2018).

35. U.S. Energy Information Administration, Annual Energy Outlook 2017, #AEO2017, January 5, 2017, available for download at https://www.eia.gov/outlooks/aeo/pdf/0383(2017).pdf (accessed October 14, 208).

36. U.S. Energy Information Administration website, available at https://www.eia.gov/tools/faqs/faq.php?id=105&t=3 (accessed September 17, 2017).

37. U.S.-Canada Power System Outage Task Force Final Report on the August 14, 2003, Blackout in the United States and Canada: Causes and Recommendations, April 2004, available for download at https://energy.gov/sites/prod/files/oeprod/DocumentsandMedia/BlackoutFinal-Web.pdf (accessed October 14, 2018).

38. American Recovery and Reinvestment Act of 2009, H. R. 1–24.

39. U.S. Department of Energy website, Office of Electricity Delivery and Energy Reliability Smart Grid website, section on "What Is the Smart Grid?" available at https://www.smartgrid.gov/the_smart_grid/operation_centers.html (accessed April 26, 2018); U.S. Department of Energy, Office of Electricity Delivery and Energy Reliability website, Grid Modernization Initiative, available at https://energy.gov/under-secretary-science-and-energy/about-grid-modernization-initiative (accessed September 16, 2017).

40. U.S. Department of Energy, Grid Modernization Multi-Year Program Plan, November 2015, available for download at https://energy.gov/sites/prod/files/2016/01/f28/Grid%20Modernization%20Multi-Year%20Program%20Plan.pdf (accessed October 14, 2018).

41. U.S. Department of Energy website, section on "Shale Gas 101," available at https://energy.gov/fe/shale-gas-101#faq0 (accessed September 17, 2017).

42. U.S. Energy Information Administration website, Natural Gas Statistics, available at https://www.eia.gov/energyexplained/index.cfm?page=natural_gas_home#tab2 (accessed October 14, 2018).

43. U.S. Energy Information Administration website, Electricity Statistics, available at https://www.eia.gov/energyexplained/index.cfm?page=electricity_home#tab2 (accessed October 14, 2018).

44. U.S. Energy Information Office website, Natural Gas Statistics, available at https://www.eia.gov/energy-explained/index.cfm?page=natural_gas_home#tab2 (accessed September 17, 2017).

45. U.S. Energy Information Administration website, section "About U.S. Natural Gas Pipelines," available at https://www.eia.gov/naturalgas/archive/analysis_publications/ngpipeline/index.html (accessed September 17, 2017).

46. Federal Energy Regulatory Commission website, Natural Gas Information, "NGA Hinshaw Pipelines," available at https://www.ferc.gov/industries/gas/gen-info/intrastate-trans/hinshaw.asp (accessed September 17, 2017); U.S. Energy Information Administration website, Delivery and Storage of Natural Gas, available at https://www.eia.gov/energyexplained/index.cfm?page=natural_gas_delivery (accessed September 17, 2017).

47. U.S. Energy Information Office website, Natural Gas Pipelines, available at https://www.eia.gov/energy-explained/index.cfm?page=natural_gas_pipelines (accessed September 17, 2017).

48. Federal Energy Regulatory Commission website, available at https://www.ferc.gov/about/ferc-does.asp (accessed September 16, 2017).

49. U.S. Department of Education, *Digest of Education Statistics*, Table 203.10. Enrollment in public elementary and secondary schools, by level and grade: Selected years, fall 1980 through fall 2025.

50. U.S. Department of Education, "Projections of Education Statistics to 2023, Forty-Second Edition," April 2016.

51. Act to Establish a National Park Service (Organic Act), 1916; National Park Service website, section on "History," available at https://www.nps.gov/aboutus/history.htm (accessed September 24, 2017); 16 USC 1.

52. National Park Service website, section on "FAQ" available at https://www.nps.gov/aboutus/faqs.htm#CP_JUMP_5057993 (accessed August 20, 2017); U.S. Research Report Office Market Outlook Q1 2017, Colliers International.

53. National Park Service website, Basic Information, section on "National Park or National Forest?" available at https://www.nps.gov/grsm/planyourvisit/np-versus-nf.htm (accessed September 30, 2017); U.S. Forest Service website, section About Us, available at https://www.fs.fed.us/about-agency (accessed September 30, 2017).

54. General Service Administration website, Federal Real Property Profile Management System, available at https://www.gsa.gov/policy-regulations/policy/real-property-policy/asset-management/federal-real-property-profile-management-system-frpp-ms (accessed December 16, 2018).

55. FY 2017 Federal Real Property Profile (FRPP) Open Data Set (most recent available), available for download at https://www.gsa.gov/policy-regulations/policy/real-property-policy/data-collection-and-reports/frpp-summary-report-library (accessed December 16, 2018).

CHAPTER 3

RECOGNIZING OPPORTUNITY

Recognizing opportunity is the first development milestone, as discussed in Chap. 1, which must occur in order for any project to be realized. Exactly when and how a development project begins, however, is difficult to define because not all projects start from the same point. For the developer, recognizing opportunity and beginning a project is a challenging and creative undertaking. Before involving allied professionals such as engineers, architects, and attorneys, the developer starts with nothing and must find an opportunity to turn into a viable development project. In order to determine exactly what scenarios constitute an opportunity without incurring sunk costs for every nonviable site considered, the developer must often make early assumptions and operate with incomplete information. This can be an uncomfortable concept for civil engineers and other process-oriented development team members, whose involvement usually begins after a project has gained at least some structure. Friction can easily result because many professionals on the development team typically only become involved with a project after the developer has already made preliminary assumptions that may affect them. For example, an engineer will usually have his/her required tasks laid out in a scope of work (SOW) or, in the absence of a SOW, will follow certain processes for technically evaluating each site. However, the engineer may consider the developer's proposed SOW to be lacking due to incomplete or insufficient information. As already stated, however, the

developer's initial assessments about a potential project often must be made on less-than-perfect information at the early stages of a project and a less robust SOW may be unavoidable.

As mentioned in Chap. 1, there are many different kinds of developers who may pursue different kinds of deals and source opportunities in different ways. This chapter will discuss the different types of potential sites and development projects that a developer may consider. It will also outline the developer's preliminary considerations at the inception of a project. These are crucial factors that influence development Milestone 1 (recognizing opportunity). Considerations at these early stages often happen before the development team is assembled and it is helpful for civil engineers and other development team members to understand how early decisions are made and how they may affect the future project as well as development team challenges. Further, appreciation for the developer's early process may help the team appreciate why subsequent information packages or SOWs are not always provided in a complete or ideal state.

3.1. TYPES OF SITES

When looking for project opportunities, sites may represent different opportunities depending on a developer's individual preferences, risk tolerance, and expertise. This section will describe some common types of development projects and associated challenges.

3.1.1. Acquisition versus Ground Lease

Purchasing land and then developing a project on the newly acquired site is perhaps the most common way for developers to undertake a project. However, it is also possible for a developer to build a building (or buildings) on a site that he/she leases but does not own. This arrangement is known as a *ground lease*. Various factors influence the decision to own or lease property for development.

Developers generally prefer to purchase the sites they develop because ownership of the underlying land gives them full control of the property. It also makes the future sale of the development easier because a single entity holds legal title to both the land and buildings. In the United States, the most complete form of private ownership of real property is *fee simple absolute*. This form of ownership, known as *freehold*, provides five key rights to the holder, which are the rights of (1) possession, (2) control, (3) exclusion, (4) use, and (5) disposition. The right of exclusion refers to the owner's right to exclude others from using the property; the right of disposition refers to the owner's right to sell or lease the property. The other rights are fairly self-explanatory, although court cases have certainly arisen over unusual or extreme circumstances for each. These five rights are often referred to collectively as the "bundle of rights" of real property ownership. They apply not only to the surface of the property but also extend to subsurface rights and air rights. It should be noted that ownership rights have practical limitations. For example, property owners cannot stop airplanes from flying above their properties. Effectively, private air rights end at the height where navigable airspace begins, as defined by the federal government.[1] Similarly, zoning ordinances, building codes, and other regulatory restrictions limit exactly what an owner can develop on a property, which impacts an owner's right to use the property.

Despite a developer's preference to own the land upon which he/she will build, the owner of a well-located piece of land may be unwilling to sell. In instances where a current landowner is not willing to sell but equally unwilling (or unable) to develop a property, the landowner and the developer may benefit from a ground lease arrangement.

A *ground lease* is a special lease in which a developer leases the right to access and build upon a parcel of land for a period of time in exchange for lease payments to the landowner. Under a ground lease, the landowner is known as the *ground lessor* (i.e., the "landlord" for the ground) and the developer is known as the *ground lessee* (i.e., the ground "tenant"). Unlike a purchase of land, under a ground lease the developer has a *leasehold* interest in the land but does not own it. However, the developer owns the building he/she builds on the land and is entitled to access it, use it, sell it, or lease it to others and collect rental payments as he/she would otherwise. Because the developer is paying to construct the building, he/she will want the longest possible term for the ground lease to make sure all the economic value has been captured from the project and the development costs are fully recovered by the collection of rental revenue. Therefore, the initial lease term—at least theoretically—reflects the expected useful economic life of the developer's project. Accordingly, the term of a ground lease may last for several decades and a 99-year term is not at all uncommon, although the landowner and developer can agree to any term length they choose.

While ground leases are a useful tool for creating opportunity and facilitating a development project when a landowner does not want to sell a site, they have several disadvantages. First, the developer must continue to make rental payments to the ground lessor (landowner) for the land for the entire term of the ground lease. These payments increase overall costs and impact a proposed project's financial returns. In some cases, the extra cost may render a potential project unfeasible. Second, the title to the building (held by the developer/ground lessee) is separate from title to the land (held by the landowner/ground lessor), which can make the buildings on a ground-leased property difficult to finance or sell to investors. This means that ground-leased sites represent less desirable opportunities in general, especially for merchant builders. Finally, at the end of the ground lease term, the developer's building typically becomes the property of the landowner, unless the ground lease contains some provision to the contrary.

The value of a ground lease can explain why a landowner may not want to sell a well-located parcel. For example, a 2-acre parcel of retail land in a suburb of the Washington D.C. metro area may be valued for purchase/sale between $1 and $2 million, depending on its most suitable use. However, the same parcel of ground may represent the following annual income streams if leased to (A) a fast food restaurant: $150,000 per year in ground lease rent, (B) a gas station: $215,000 per year in ground lease rent, or (C) a national pharmacy chain: $325,000 per year in ground lease rent. The total rent collected on such a parcel over the period of a 50-year ground lease would total, respectively: $7.5 million, $10.75 million, or $16.25 million. From the landowner's perspective, the ground lease revenue stream combined with the potential ability to obtain a loan against the ground-leased value of the land can make ground leasing a much more favorable option than a sale.

Considerations for Developers. Real estate markets change over time and conditions are likely to be different 20 years into a ground lease than they were at its inception. Accordingly, a ground lessor (the landowner) may require rent reviews, rental increases, or reappraisals throughout the decades-long term of a ground lease. Any increases in the underlying ground rent will impact the overall financial returns of the project, so developers must consider when and how rent increases will occur in order to account for them in advance. Ideally, the developer will structure leases for the newly constructed building such that any increases in ground lease rent can be "passed through" to the future tenants of the building.

In projects where the developer has purchased a freehold interest in a site, he/she can also create a ground lease opportunity. In these cases, the developer is the ground lessor and collects ground rent on a small portion of the larger project site. This typically occurs in the context of retail projects where, in addition to building and leasing the majority of the property, a developer may also designate smaller areas within the site to ground lease to specialty retailers. These areas are called *pad sites*, or *outparcels*, and are designed to accommodate small, freestanding buildings. Pad sites will be discussed further in subsequent sections of this chapter. In a redevelopment scenario, some developers actually purchase properties with the specific intent to capture opportunity by creating ground leases. Such opportunities are often recognized in an existing shopping center that may include "excess" land, either used for landscaping or extra parking beyond what the current market or zoning regulations demand. Depending on the density allowed under the zoning, these "excess" areas may be repurposed as pad sites and ground leased by the developer.

Considerations for the Design Team. Ground leases can be complex undertakings from the architectural/engineering perspective. In a ground lease scenario, the engineer (and/or architect) responsible for site layout effectively has two clients: the developer and the landowner. Both may have their own requirements with respect to how a site is developed and no layout will be approved unless it satisfies both parties. Thus, while the engineer is officially part of the developer's team, he/she must also understand the landowner's concerns in order to incorporate them into the site design. This can lead to challenging situations in which engineers are caught between their developer-clients and the landowner, particularly when a technical evaluation of the site reveals that what either the landowner or developer (or both) wants isn't technically feasible. The responsibility may fall to the engineer to coordinate between the developer and landowner to find a solution that satisfies all parties and still makes economic sense for the developer's project.

Further, if a ground lease does not encompass an entire site, the larger site will often be separated with "lease lines." These are distinct from the parcel's boundaries and do not create a new, individual site; rather, they delineate the portion of the site subject to the ground lease. The creation of lease lines may have implications for utility requirements, such as water and sewer utilities that now must be brought into the ground-leased area. Additionally, private easements between the developer and landowner are often required so that common areas can be shared; these are often referred to as *reciprocal easement agreements*.

In ground leases involving pad sites within a development project (where the developer is the ground lessor), it is possible that multiple design teams will have to work together during the late stages of design. For example, a shopping center may have been designed and approved—and even be under construction—before a bank tenant signs a ground lease for a pad site. Regardless of what previous layout or design decisions have been made for the shopping center, the bank tenant will have its own corporate requirements and expectations about site design and will often engage its own design team. The developer's engineering and architecture team members must work with the bank design team to coordinate or resolve competing needs/wants, such as a change to the site access or traffic drive-lanes surrounding the pad site.

Considerations for Attorneys. In either an acquisition or a ground lease, attorneys are concerned with ensuring that the contract between the developer and the landowner is thorough and envisions sufficient protection and remedies for the developer under a range of potential scenarios. Acquisition of property is usually facilitated by a Purchase and Sale Agreement or through the use of an Option Contract. These contract vehicles will be discussed further in Chap. 4. With respect to a ground lease, attorneys will help the developer assess all relevant contract terms, such as those that govern rent increases, subletting and assignability, and cure of default. In instances where a rent increase is tied to an appraisal, future disputes can arise with respect to how the land is to be valued unless the methodology is clearly stated in the ground lease. Similarly, the right for the developer to assign the ground lease or sublet the premises to credit-worthy assignees/subtenants is not guaranteed unless permitted by the ground lease. The ground lessor may require completion of any improvements before allowing assignment/subletting and may also want to restrict or prohibit certain uses as a condition of allowing an assignment/sublet.

Attorneys must coordinate with the developer and lender to ensure that lender concerns are successfully incorporated into the ground lease agreement. For example, the ground lease may be "subordinated" or "unsubordinated." *Subordination* refers to the priority of claims or ownership interest in an asset. In a subordinated ground lease, the ground lessor (landowner) agrees that its ownership interest is subordinate to (lower priority than) the lender's interest. This effectively puts the ground lessor in the position of a junior lender that is pledging the land as collateral for the loan on the improvements. The ground lessor may agree to this if the improvements to be built on the ground-leased land will add value to the ground lessor's other interests. For example, when a ground-leased pad site in a shopping center is built with a drive-through pharmacy or restaurant, this adds value to the rest of the shopping center. The ground lessor may also be able to negotiate higher rent for the ground-leased site by subordinating if doing so helps facilitate the developer's financing of the improvements. In an *unsubordinated* ground lease, on the other hand, the ground lessor retains a "first place" claim position. In such cases, if the developer defaults on his/her financing, the lender cannot take the underlying land. Ground rent is typically lower in unsubordinated ground leases because risk is transferred to the developer. In either case, if the developer is financing the improvements, then a lender will be involved and the lender's interests impact the terms of the ground lease. For example, if there are restrictions on assignment and subletting of a ground lease, the lender will want to be exempt from such restrictions. The lender may also want the right to cure any developer default under the ground lease and will want all amendments and termination of the ground lease to be subject to lender consent. Attorneys will help their developer-clients navigate the related contract clauses.

Considerations for Lenders. Typical loans related to the purchase, development, and ownership of commercial property are collateralized by the underlying land as well as the buildings being developed on the site. This allows the lender to foreclose on the entire property in the event the developer defaults on the loan. However, as previously described, ground leases introduce new considerations into financing development projects because the developer does not own the land, but merely leases it. This means the lender's relationship is not directly with the landowner and, depending on whether or not the ground lease is subordinated or unsubordinated, the lender may not be able to foreclose in the event the developer defaults. In unsubordinated ground leases, it is the developer's leasehold interest—not the land itself—that must serve as the lender's collateral. Thus, unless the landowner agrees to subordinate its interest to the bank, the lender will have its own requirements for the ground lease in order to be satisfied that sufficient value exists to justify the loan. For example, the lender may limit the loan size to only the improved value of the building to be constructed. Also, lenders may require that the length of the ground lease term extend for a specified period of time (e.g., at least 20 years) beyond the mortgage. This ensures that the loan will not mature before a ground lease expires and gives the lender some assurance that, if it must foreclose on the building the developer constructs (but not the underlying land), the remaining lease term will provide sufficient value to allow the lender to sell the lease and recover the debt.

3.1.2. Assemblage versus Subdivision

Not all available sites are optimally useable in their existing configurations. A site may be too large or too small to accommodate a particular development project. In such instances, the developer can pursue either an assemblage or subdivision. In undertaking either, the developer will need support from his/her attorney and civil engineer.

A well-located site might simply be too small to be viable. In such cases, designing a smaller building to fit the site is not necessarily a solution if the total rent that can be charged on the smaller building is not sufficient to cover acquisition, development, and operating costs. However, if an adjacent property is available for sale, the developer can acquire both sites and legally retitle them into a single site large enough to accommodate an economically viable project. This process is called *assemblage* or, sometimes, *consolidation*. Assemblages are not limited to acquiring the occasional extra piece of land for an individual project. Developers can create entirely new sites—and new opportunities—by acquiring several parcels of land and retitling them into a single, larger piece of property. It should be noted that assemblages

can be costly and time consuming. Instead of negotiating with a single owner, the developer must negotiate with several owners who each may want different prices and terms. Further, not all of the land in the desired assemblage area may be for sale, so the developer will likely have to pay a premium for some parcels in order to incentivize a sale. In some cases, owners are simply unwilling to sell, which can jeopardize an assemblage project. While public sector developers can force a sale through the use of eminent domain, private developers cannot. Thus, if a current owner refuses to sell to accommodate a private sector development assemblage, also known as "holding out," it can create inefficient results for the final project. In these situations, the civil engineer and architect will need to advise the developer on how best to program the site and building given the limitations caused by the hold out.

Conversely, instead of finding several smaller parcels, a developer may find a large tract of land that represents opportunity but is too large in its current form. In such cases, the developer may take the larger parcel and go through the formal process of dividing it into individual lots. This process is called *subdivision* and the legal mechanism by which individual lots are created is called *platting*. The newly subdivided lots are each recorded as separate parcels with separate title deeds. This process is not limited to large, rural parcels and can also be done with smaller, infill parcels. There are several reasons a developer might subdivide a property, such as to build and sell individual homes or to sell several lots and use the proceeds to offset the cost of developing parcels retained by the developer. A developer may also use subdivision to break a large site into many separate project phases, which can then each be financed, built, and leased or sold individually. This allows the developer to proceed in stages and engage with a much larger overall project than he/she might otherwise have the capacity to undertake. Figure 3.1 provides an example of a subdivided site.

While some developers will engage in either assemblage or subdivision as a necessary process to facilitate a building project, not all firms involved with development have an interest in completing the entire process. Some specialty developers focus

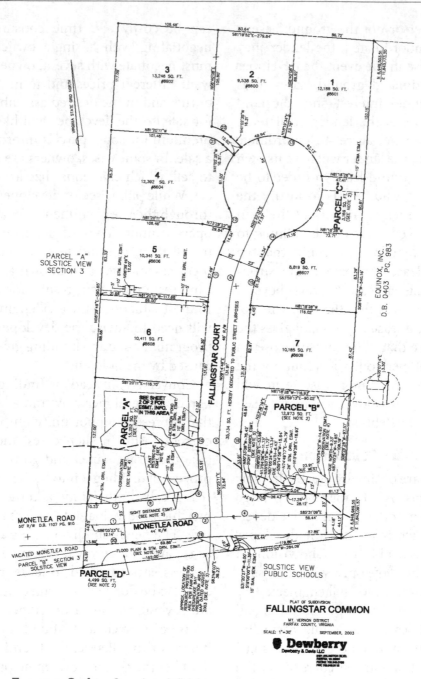

FIGURE 3.1 Sample subdivision.

exclusively on one portion of the process and cultivate niche competencies in that specific area. They then use their abilities to add value at their stage of the process before reselling a property to another developer. Developers that specialize in either creating assemblages or subdivisions are examples of such specialty developers. A developer specializing in assemblages can sell the new, larger parcel at a premium to a future developer who is saved from

the time and risk of having to create the assemblage. Conversely, a subdivision specialist developer may take a large piece of land and go through the formal process of dividing it into individual lots so that houses can be built on each. Such a site can also be sold at a premium, for example, to a developer specializing in homebuilding who is willing to pay more for a site that is already subdivided and ready for home construction.

It is possible for both assemblage and subdivision to occur across the same project at different points in time. For example, a "master developer" can assemble land; plan an entire project, including the various product types and project components; go through the regulatory process to obtain approval for the development; and then subdivide the assembled land into new properties that match the different asset classes or phases in the master plan. The newly subdivided parcels can then be retained by the developer or sold to developers specializing in the different asset classes. For example, a large development project may include a hotel site that will be subdivided into an independent lot and then sold to a hotel developer or operator.

Considerations for Developers. When subdividing land, particularly with the intent to sell, developers must be sure that there is market demand for the resulting parcels. When undertaking an assemblage, developers should, whenever possible, get key parcels under contract first. Additional parcels can then be added to enhance the project but will not jeopardize it if the developer is unable to acquire them. Developers should be aware that neighboring site owners may be incentivized to raise their sale prices or hold out after learning that a developer is buying surrounding parcels. Thus, when approaching different sellers, developers may attempt to negotiate under different corporate names or entities in order to keep their activities private and protect the integrity of the project.

Considerations for the Design Team. In both assemblages and subdivisions, the design team plays an essential role. In order for a developer to complete either process, municipalities require an engineer/surveyor to create a formal plat showing the new proposed parcel lines. When consolidating or subdividing land, these new parcel lines may impact design requirements established by jurisdictions for building setbacks, utility services, public road access, etc. The shape and size of a new parcel may also be regulated by the jurisdiction.

In large developments, subdivisions may require additional agreements and easements. For example, in the subdivision plat, parking for a building may technically be located on a separate lot for various logistical reasons. Even when both parcels are clearly part of the same larger development, subdivision requirements often treat each parcel as a separate entity that cannot rely on adjacent infrastructure or parking. Civil engineers and other design team members need to be able to advise their developer-clients about requirements for additional easements, conditions, or approvals from the jurisdiction that may impact the subdivision project.

In addition to regulatory concerns, value can be gained or lost based on how a site is subdivided. The design team should be involved in the early project concept stages to help the developer determine the best way to partition land to ensure the marketability of the resulting parcels.

Considerations for Attorneys. Assemblages can be complex because the developer must come to agreeable terms with each individual parcel owner (sellers), while collectively negotiating timing and contingencies. For example, while individual contract terms, such as purchase price, may differ depending on parcel size and other factors, there may be common terms (or even a master agreement) providing that the developer is not obligated to acquire each site unless individual contracts are executed for a certain number of parcels or for particular parcels that are critical to the assemblage. Different parcel owners may insist on different terms, conditions, and closing dates. However, the contracts must be coordinated for the entire assemblage so that different terms for different parcels do not adversely impact the larger project. This means the underlying contract terms themselves often need to be different when the attorney prepares contracts for an assemblage as opposed to the purchase of a single parcel. Contracts used for an assemblage should envision what will happen for a particular parcel/acquisition if the developer fails to come to terms with other sellers and is unable to acquire other key parcels. Various contingencies and conditions to closing should be incorporated into each contract so that the developer is not forced by contract to continue to close on parcels that have value only if the assemblage is successful. If possible, the developer and attorney will attempt to negotiate a partial refund of the deposit(s) in the event the developer fails to get all the necessary parcels under contract by a certain date. However, contract terms that protect a developer undertaking an assemblage can be difficult to justify because the seller(s) may have little interest in offering such terms to the developer. From the seller perspective, terms that protect the developer from the obligation to purchase the property are undesirable as they jeopardize the certainty of the sale. For this reason, option contracts are often used to facilitate assemblages. As explained in Chap. 1, an option contract gives the developer the unilateral right, or option, to purchase a particular site during a specified period of time if he/she chooses to do so. Options will be discussed further in Chap. 4.

Considerations for Lenders. In some instances, financing may be needed for acquiring parcels for an assemblage. However, the developer will generally only approach lenders for a loan after an assemblage or platting is completed. The lender will require proof of (re)zoning or approval of subdivision plat prior to loan approval or releasing funds.

3.1.3. Greenfield versus Brownfield

A site that has never been previously developed is called a *greenfield site*. These sites have no, or extremely minor, existing man-made improvements. Greenfield sites represent good opportunities for developers because, by virtue of having never been previously built-upon, they do not require any demolition and are less likely to have any prior contamination. These sites may also have heavy tree and vegetation cover, which can contribute to aesthetic natural landscaping, particularly for a new residential community. However, greenfield sites also have disadvantages. They primarily exist in suburban or rural areas and, as such, may be located a considerable distance away from existing utility capital improvements. As discussed in Chap. 2, development costs are increased by the need to run the necessary utility connections to a site. Also, a large portion of the natural cover on a greenfield site may need to be preserved as per any prevailing tree ordinances and zoning open space requirements. Depending on the type of the existing natural features, wetland preservation

regulations may also apply. These types of preservation requirements, while environmentally important, can impede efficient site layout and be costly to accommodate.

In contrast, a *brownfield site* is one that has been previously been developed for another use. These sites are usually located in urban areas or near-in suburbs, sometimes also called *peri-urban* areas. As such, brownfield sites may be well located in established markets. They may also be located in an economically depressed area but could be of interest to a developer looking to capture the reduced cost of a distressed property and create value through redevelopment and the economic revitalization of the area. One of the main challenges with brownfield sites is the need for demolition of existing structures and the potential for existing contamination, which the developer must pay to remediate.

Considerations for Developers. In both greenfield and brownfield projects, the developer must account for the removal of dirt and debris in both the construction schedule and budget. However, brownfield sites that involve heavy demolition also require the demolished material be removed, which adds additional cost to the project. In the course of completing site work and/or excavation in both greenfield and brownfield projects, developers can encounter challenges if archeologically significant discoveries are made, such as the uncovering of graves, artifacts, or historic structures. This is often a higher risk in urban redevelopment projects, where human occupation has left historical traces for decades or even centuries.

Brownfield sites pose several unique challenges. Potential contamination is always a concern developers must be aware of with brownfield sites. The cost of remediation is prohibitive for many developers and there is always a risk that unknown contamination will be discovered even after initial tests have been completed. Although they may be vacant lots where previous buildings have already been demolished, brownfield sites often have both above and below ground existing structures that need to be demolished and/or removed in order to accommodate new development. Demolition and material removal can be complicated in urban environments where dump trucks and other equipment may not have adequate room to maneuver without blocking streets. Further, urban removal necessitates trucks engaging with traffic, which can reduce timeline efficiencies and increase costs. In some instances, the existing structures may have historic value and require full or partial preservation. Both demolition and historic preservation obligations can create substantial increases in project cost and time for the developer.

Considerations for the Design Team. Greenfield sites may have unmapped environmental areas that need to be identified in the early stages of design. The presence of wetlands, floodplains, historic areas, and other environmentally relevant characteristics can introduce constraints that impact site layout, overall project design, and development requirements. These must be accounted for by the design team. For example, additional provisions may be necessary to accommodate environmental protection measures, such as stream restoration or erosion and sediment control, which are often required by the municipality or the U.S. Army Corps of Engineers as part of the approval process.

Brownfield sites require environmental investigations to determine if any contamination is present and remediation required. For some types of brownfield sites, such as a former landfill, there may be regulatory limitations restricting new development on the site until a certain amount of time has passed. Additionally, there may also be mandatory monitoring and structural design requirements that must be accommodated. Each of these measures contributes to the project's cost.

From a civil engineering perspective, brownfield sites benefit from existing infrastructure, such as a water or sewer connection. However, assessing the condition of the existing infrastructure is often difficult and may require significant replacement and/or repairs. The civil engineer and other design team members must make additional provisions to account for demolition, phasing of work, assessment and repairs as may prove necessary, all of which adds to the overall project cost.

Considerations for Attorneys. While attorneys are always cognizant of environmental risks that must be addressed, these can often be of greater concern in brownfield projects. To ensure the developer does not inherit liability for previous contamination, the attorney will make sure the purchase contract contains an environmental indemnification provision pursuant to which risk for liability for remediation is allocated. Further, the contract will stipulate a "study period" in which the developer can conduct due diligence to evaluate the site in detail, including environmental testing and analysis. At a minimum, for both greenfield and brownfield sites, the developer should obtain a Phase I Environmental Site Assessment (ESA) report. Depending on the results of the ESA, a more comprehensive Phase II assessment may be required. Due diligence, including Phase I and II ESAs will be discussed further in Chap. 5. Note that the developer may be able to offset risk by purchasing insurance products that underwrite redevelopment deals and provide liability coverage and cleanup cost coverage for brownfield sites.

Considerations for Lenders. Lenders require environmental assessment reports to be completed and submitted as part of a loan application. While a lender normally only requires a Phase I ESA report for greenfield sites, a more comprehensive Phase II ESA will almost certainly be required before a loan is approved for a brownfield site development project. The lender is also likely to require additional studies. Depending on the scope of a brownfield project and the level of known contamination, the developer may need the approval and help of federal, state, and local government agencies as well as tax credits or other additional, specialized financing.

3.1.4. Redevelopment

As the previous section described, not all development projects occur on greenfield sites. When a developer pursues a redevelopment opportunity on a brownfield site, there may be existing buildings on the property. In a redevelopment scenario, any existing buildings will be either demolished or rehabilitated. Rehabilitation projects are sometimes called *retrofits* but are more often referred to as "rehabs." The decision to retrofit or demolish a building is made based on several factors, including the condition of the existing building, zoning considerations, market demand for a specific use, costs, and time. If the developer chooses to pursue a rehab project, the degree of rehabilitation can vary between minor improvements to a complete renovation of the entire building.

For example, a developer may purchase an older building that is in poor condition but which is structurally sound. Because of the building's age and condition, it most likely commands lower rents and may have a higher vacancy rate than newer buildings in the same area. The building also likely has older, inefficient systems that are costly to operate. As a result of low rents and high operating costs, the building's net operating income (NOI), or *cash flow*, is lower than that of a newer building. The purchase price should be lower as a result of these financial circumstances. Despite perhaps being questionable investments from an operating cash flow perspective, when such buildings are well-located, they may represent a *repositioning* opportunity for a developer. By repairing or upgrading older systems as well as making strategic cosmetic improvements to the building facade and common spaces, such as entrances, lobbies, and hallways, the developer can give the property a new image and "position" in the market. The developer's goal with these upgrades is to be able to raise rents while also attracting new tenants. He/she may also be able to reduce operating costs through certain upgrades, such as adding new, energy efficient windows or upgrading the building's heating and cooling systems. This type of retrofit project is also referred to as a *value-add* opportunity because the developer essentially creates new economic value by increasing rental income and reducing operating costs, which increases the property's NOI and overall value.

Note that not all repositioning/value-add opportunities are tied to rehabilitating older buildings; a newer but poorly managed, under-performing asset could also represent a repositioning opportunity for a developer. Equally, lack of market demand for a particular use can be a particularly compelling incentive to retrofit and reposition properties for new uses, a process known as *adaptive reuse*. This is often seen in office and warehouse conversions (usually to new multifamily and/or retail uses).

Most buildings have a useful life of between 30 and 50 years, although some can remain functional for much longer. At some point in the building lifecycle, current technology, safety or other regulations, and building systems will have changed significantly enough that the building can no longer be used effectively as per new standards. Although the original building structure, facade, and systems may still be in place, they may be severely degraded and no longer able to be upgraded due to physical space constraints, extreme cost, or updated building code requirements. At this point the building is said to have reached a point of *functional obsolescence*. It may be worth noting that multifamily apartments and commercial buildings also have a depreciable economic life, 27.5 years and 39 years, respectively, after which point tax depreciation can no longer be taken for the asset.[2] However, a building's economic life and its functional/useful life are often not the same.

Once a building is functionally obsolete, the developer must decide if the time and cost to rehabilitate it are justified by sufficient returns or if demolishing the building and redeveloping the site is more appropriate. Iconic buildings and those subject to historic preservation requirements will usually be rehabilitated despite cost. In these cases, the rehabilitations can be extreme and may be more appropriately called redevelopment. Whole building systems can be replaced, new amenities can be added, landscaping can be redone, and tenant space can be improved. In some cases only the building facade is retained and the rest of the structure and interior systems are entirely removed. Note that the decision to demolish a building can be made in advance of the point of full functional obsolescence as there are certain advantages to demolition and redevelopment over rehabilitation.

Considerations for Developers. When contemplating a rehabilitation project, a developer must weigh the project's financial benefits against his/her personal risk threshold for unexpected circumstances. Even when the development team has extensive rehabilitation experience, no degree of inspection can reveal every hidden detail of an older building. Unpleasant surprises can be significant in scope and costly to remediate, such as the discovery of structural damage or asbestos. Some problems may be less extreme but still difficult to resolve. For example, the developer may discover that new systems cannot be installed as expected because of physical constraints. A simplified example of this is if a newer elevator model is too large to be accommodated in a smaller, existing elevator shaft. If the scope of the rehabilitation project is small, the noncompliant elevator may be "grandfathered" in and the municipality will permit the developer to leave the existing elevator in place.

If the scope of the rehabilitation is extensive, however, it may trigger the need for the building to become compliant with current standards and codes. In this case, the cost to replace the elevator will include not only the new machinery and equipment, but also the cost of expanding the elevator shaft. In addition to increasing project costs, potentially dramatically, installing a larger elevator(s) will also reduce the building's rentable floor space and may impact projected rental revenue. An elevator upgrade is only one example of the type of challenge posed by rehabilitation projects. The developer may encounter a similar scenario with other code-dependent elements such as life safety systems, bathrooms and access that are compliant with the Americans with Disabilities Act (ADA), parking requirements, building materials, and other design components. If consulted in advance, the design team and attorney can usually help make the developer aware of the types of components that will require upgrading before the site/building is purchased. However, if a condition is discovered during redevelopment that can only be remediated by unexpectedly expanding the project scope, certain components may no longer be "grandfathered" and the developer and development team may be unprepared for the need to accommodate new upgrades.

Considerations for the Design Team. As already mentioned, different levels of redevelopment may trigger different design requirements. A building undergoing a minor rehab may be grandfathered under old/original site and building design requirements, while a more extensive redevelopment featuring new construction may trigger the imposition of new requirements. For example, regulations in place when the site was originally developed might not have required stormwater management systems, but new regulations demand that new stormwater facilities are constructed. These systems can be costly to incorporate and often require a large amount of space, so any such regulatory changes should be considered early in the design process. Similarly, new jurisdictional requirements that may be enforced during redevelopment can pertain to parking availability, ADA access, road access, open spaces, and a range of other site or building features.

In some cases, a redevelopment project can make design easier because existing land use and utility locations can be retained. However, in many cases, infill redevelopment introduces additional and/or complex variables that need to be considered in design and construction. Assessment of existing infrastructure is often difficult and record drawings are not always available or reliable. This makes it challenging for the civil engineer and design team to determine the locations of utilities and easements, pavement section/condition, original utility demands for the site, and other relevant development conditions. Easement records may be used to identify utility locations but subsurface utility exploration (SUE) is often required, which can add time and cost to a project. Even with proper exploration/investigation, utility locations may be approximate and exact conditions cannot always be determined. In most redevelopments, it is highly likely that unknown conditions will be discovered during construction, such as irrigation lines, small site electrical lines, or design deviations that occurred during the original construction on the site. The design team should coordinate with utility providers to determine locations, research old records for guidance, and disclose any assumptions to the owner, contractor, and reviewing agencies.

Many redevelopment projects, such as those in campus or hospital settings, will require additional coordination and phasing. For example, on a large site simultaneous conditions may include (1) old buildings being demolished, (2) some existing buildings continuing to be occupied and operating, and (3) new buildings under construction. Accommodating this level of concurrent site activity often requires the engineering/design team to plan for the temporary relocation of utilities and vehicular and pedestrian routes on the site. Projects of this scope are extremely complex, lengthy, and expensive.

Considerations for Attorneys. Given that redevelopment involves a previous/existing development project, it is especially important for the developer and attorney to fully investigate the possibility of preexisting legal constraints. A variety of factors must be examined before undertaking redevelopment, including:

- Any encumbrances by liens

- Recorded covenants, conditions, and restrictions affecting the property

- Prior rezoning or other prior legislative action affecting the property and any related proffers or special use conditions

- Easements affecting the property

A title review is essential to discovering these conditions. Title reviews and easements, liens, and covenants will be discussed in greater detail in Chap. 5.

Legal issues also can arise when a redevelopment project is located in an area the municipality has designated as a revitalization area. This may have various compliance implications as well as positive and/or negative effects on a project. For example, if the developer will benefit from tax abatement incentives designed to encourage redevelopment in the area, it may reduce overall project costs and have a positive financial impact on the project. Conversely, if the redevelopment occurs in an area that is a designated historical district or other overlay zone, the historic designation or particular overlay district may impose strict, additional requirements that the developer maintain certain building features, use certain materials or finishes, or even restrict allowable uses. Such requirements can add time and cost to a project, which can have a negative impact on financial returns. In some cases, the burden imposed additional requirements can render a project infeasible.

While any urban development requires mitigating the impact on neighboring properties,

a redevelopment project that requires extensive demolition carries additional risks with respect to damaging and/or disrupting adjacent buildings. Further, older buildings may have asbestos that requires special remediation measures. While there are many "form" construction contracts utilized in the industry, all contracts are negotiable, and the construction contract must be carefully negotiated with respect to allocation of such risks. For example, if the parties intend for liability for construction practices to be borne by the general contractor rather than the developer, this must be stated in the contract.

Considerations for Lenders. In the case of a redevelopment, the size of the loan will be based on not only the projected future cash flow of the project but on the "as-renovated" value of the property as well. Further, if the building is in an historic area, the historical preservation designations may increase the project cost and/or timeline. This has implications for the size of loan and timing of repayment, both of which impact the lender. The lender will want assurance that the development team includes qualified contractors that have experience with historical properties.

3.2. PROPERTY TYPES

Developers may specialize in a particular product type, such as residential, retail, office, or industrial. There are several similarities across each of these product types. For example, as outlined in Chap. 2, buildings can all be described by the same height classifications: low-rise or "garden," mid-rise, and high-rise. However, despite such similarities, each of the different product types represents different types of opportunities and challenges for the developer and development team. This section will briefly describe each product type and highlight special areas of consideration for developers, civil engineers, and other members of the development team. Examples of different site layouts will be provided using the sample site introduced in Chap. 2, Atherton Road (Site A).

3.2.1. Residential

There are two main kinds of residential development projects: single-family and multifamily. Single-family residential units may be either detached from one another, as with freestanding homes, or attached by a common wall, such as in duplexes or townhomes. Two key components of single-family residential properties are that each home (1) has its own entrance and (2) is placed on its own individual lot, which may include a driveway and yard. In contrast, multifamily residences are characterized by four or more units in a single building on a single lot. Common open space is shared and units usually share exterior access through one or more common entrances. Multifamily developments may include shared recreational amenities such as pools, parks, or clubhouses. Whereas homes in single-family development projects are usually sold, multifamily properties can be offered either for sale or lease. In the United States, when the units in a multifamily building are sold to individual owners, the units are referred to as condominiums. When individual units are leased and ownership of the entire building is retained by the developer (or a future investor), the units are referred to as apartments.

Developers earn income from residential projects either through sales or, in the case of apartments, from rental income and the possible future sale of the entire property to an investor. Market demand is important for the success of both types of projects, although demand for rental apartments tends to be countercyclical to demand for home purchases. In other words, when the local economic development is strong, salaries are growing, and interest rates are low, people are more likely to buy homes. This can increase vacancy in multifamily properties, especially in older properties with fewer amenities. When market demand favors home buying, multifamily owners may have to reduce rents to remain competitive and attract tenants remaining in the rental market. Conversely, when the economy is weak, jobs are being cut, and interest rates are high, people are more likely to rent, particularly in worst-case scenarios where they may have lost a single-family home through foreclosure. Under these conditions, apartment communities can raise rents and still enjoy lower overall vacancy, but homebuilders may experience slow or stagnant sales.

Considerations for single-family developers include the time and cost to subdivide land, evaluating the need to incorporate neighborhood amenities such as a pool, determining what type/size of homes to offer, and the need to manage the total cost per home relative to the achievable sales price. If the development project includes a neighborhood retail component, such as a dry cleaner, day care, or grocer, the developer will also be concerned about finding tenants for the retail space. Multifamily developers have similar concerns about incorporating amenity packages (club rooms, private gyms, rooftop terraces, etc) and *unit mix*, which refers to the types and sizes of units included in the project. In urban and peri-urban environments, a multifamily developer may include (or may be required to include as part of the project approvals) retail units at street level. In this case, the developer must also find retail tenants. Unlike single-family residential developers and condominium developers, who turn management of the community over to a homeowners' association, apartment developers must manage leasing and building operations after the project is complete. This can be done internally by the developer's firm or by hiring an external property management and leasing firm. The cost of apartment management must be included as part of the project's operating expenses.

Figure 3.2 shows a sample layout for a single-family townhome community development on the Atherton Road site.

Considerations for Developers. Developers must accurately gauge the future market demand for residential units they intend to sell or rent once their project is completed. The price that each unit can command, whether in rent or sale, must be considered at the beginning stages of the project and guide decisions as early as negotiating the purchase price for the land. If the developer pays too much for the site, it will increase the cost-per-unit of the resulting residential units and, in turn, increase the amount the developer must achieve in the sale or rent of each unit. If the cost is too high, there will be little or no interest from buyers/renters. Developers must also consider the cost of including community amenities versus the return on the investment needed to create them. A community, whether for lease or for sale, might be more attractive with a clubhouse and pool, but the

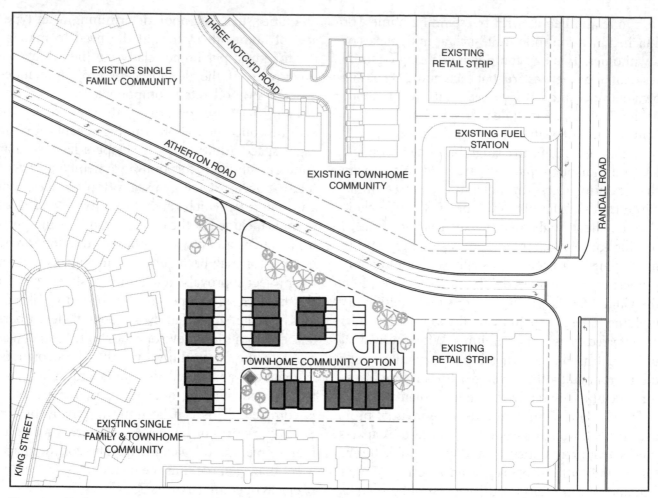

FIGURE 3.2 Atherton Road: sample townhome layout.

cost to install and maintain such amenities may not be justified by the projected sale/rental prices.

Planning for amenities, fixture quality, and other project decisions can be particularly challenging because a developer cannot simply focus on balancing his/her own costs with project returns; rather, the developer must also consider how his/her own project decisions measure against what competing development projects may provide. For example, if a rival developer has lower project costs (perhaps due to a lower purchase price of land or lower cost of debt and/or equity), he/she may be able to install amenities such as a pool and clubhouse while still maintaining market prices. If both projects deliver together at substantially similar market-driven prices, tenants/buyers will naturally gravitate to the competing project that has more amenities.

Thus, the developer needs to conduct different market studies to evaluate competing projects and market demand. Components of market research will be discussed in Chaps. 4 and 5.

Considerations for the Design Team. "Single-family attached," "duplex," "townhome," "multi-family," "apartment," and "condominium" are all common terms used to describe different types of residential properties. However, building codes may use a particular definition for each term while a jurisdiction may use another definition for the same term. The development team's civil engineers should verify definitions of each term in order to ensure the project being designed is compliant with all applicable definitions. Definitional differences may not always be readily apparent. For example, an affordable housing development may appear to be designed as single-family attached townhomes, but may in fact be considered multifamily because units exist on a single lot and share utilities.

Single-family residential projects often involve a developer and one or more homebuilders. The developer's engineer will likely need to work with all parties involved, which leads to challenges of navigating the needs of multiple "clients," as was described in the previous section on ground leases. Single-family detached projects are based on a subdivision plat, which delineates the individual lots for each home. When creating the subdivision plat, the design team must accommodate jurisdictional requirements and developer decisions about lot sizes and community amenities in order to determine the number of units that can fit within a proposed community. Lot sizes can range from under 10,000 square feet to multiple acres. Larger developments, such as those with 500 or more homes, can support community centers and pools as part of the project, but such amenities are usually not included in smaller developments. Many jurisdictions place restrictions on stormwater management systems that can be used, often prohibiting wet ponds because of hazard considerations. In general, the stormwater management is covered through systems located in common areas of a community, and not treated within the individual lots. In early design phases, the engineer may use home templates for an approximate size and location of the homes. In later phases of design, when the future homeowner has chosen a specific lot and home style, the engineer will create detailed lot grading plans to finalize the house location and lot grading.

Single-family attached projects, such as townhouses, allow for a greater density of homes within a development. Increased density can introduce design challenges for accommodating utilities because each home still requires a separate utility connection despite space constraints. Additionally, the rows, or "sticks," of attached homes may range from a span of two units (duplex) to upward of 10 units. This effectively creates a large building mass, which should ideally be designed to have a common floor elevation to minimize construction challenges and cost. However, it can be difficult to maintain a constant elevation across multiple connected homes and grading for each unit must be coordinated to account for how each adjacent home is impacted.

Multifamily developments tend to feature large buildings and, as such, require greater structural scrutiny and consideration for vertical components and systems coordination than other residential product types. Accommodating adequate parking on the site can be a challenge, as the density of multifamily units will typically require a larger number of parking spaces yet the cost for underground or structured parking (as required by small lots) is significantly higher than surface parking. The developer must predetermine if a multifamily community will be rental apartments or condominiums because converting from one to the other is often difficult and requires additional plats and design considerations.

Considerations for Attorneys. Most new for-sale residential communities, including both single-family and multifamily condominium developments, are managed by a "common interest community association" whose members are comprised of the property owners. These are usually either homeowners' associations (HOAs) or property owners' associations (POAs). An HOA or POA is a legally cognizable entity that is governed by state law and is formed by filing certain documents, known as *articles of incorporation*, with the state. Some states also require registration with a common interest community board and annual filings from the organization thereafter. The developer's attorney will help form the HOA/POA entity and bring it into legal existence before all the homes are sold. Typically, the developer retains control of the HOA/POA board until a certain number of homes are sold and then relinquishes control of the association to its members.

The attorney also works with the developer to establish *covenants, conditions, and restrictions* (CCRs) that will govern the lots/units (unless excluded). CCRs are usually described in a document known as a "Declaration of Covenants, Conditions, and Restrictions" in which the developer is known as the *declarant*. Typical CCRs help maintain the integrity of the community by establishing what colors homes can be painted or dictating certain minimum standards of acceptable property maintenance. However, they can cover virtually anything including what types of landscaping are allowed and whether or not the community allows overnight visitor street parking (if the streets are private). Additionally, the recorded covenants may or may not contain restrictions or prohibitions on homeowners renting their properties or use for home businesses. Local zoning ordinances may also address these issues and state governments and municipalities are increasingly looking into regulating short-term rentals due to the popularity of homeowners renting out their dwellings to travelers (such as with AirBnB). The developer's attorney will play a key role in helping consider the legal requirements and consequences of decisions incorporated into the CCRs. The CCRs established by the developer are officially recorded in the land records and are deemed to "run with the land," meaning they will apply to all future purchasers/owners.

Condominium developments will also have an HOA or POA, which is usually referred to as a "Condo Association." However, condominiums require additional regulation because of their unique nature and, as a result, also require an additional set of legal documents. The condominium structure itself is formed by filing a "Declaration of Condominium" with the state. State law provides a detailed regime for condominiums that is supported by an extensive legal framework that can vary from state to state. The developer's attorney will help navigate the applicable legal requirements and generate the necessary documents. Under a condominium structure, property owners have exclusive ownership only in their unit, which will have an individual title. However, all unit-owners jointly own shared areas, such as hallways, lobbies, landscaped areas, and driveways. These parts of the development are known as *general common elements*. Condominiums may also include *limited common elements*, which are those that exist in the joint space but only benefit individual unit owners, such as dedicated parking spaces or balconies.

Considerations for Lenders. In the case of both single-family homes and multifamily developments (whether for sale or lease), the developer will need to demonstrate that sufficient market demand for similar housing exists to support the additional supply of units being built. Thus, market projections in the loan package will need to be robust and withstand lender scrutiny. As an additional measure of guarantee that demand will be as-projected, the lender may stipulate presale requirements, especially for large-scale single-family developments. Under such conditions, lenders require that a certain percentage of units are sold before additional loan funds are released to continue financing construction of additional units. The lender may also want a detailed "takedown schedule" that outlines when additional lots will be transferred between the land developer and the homebuilder (if the roles are separate). For multifamily apartment developments, the lender may want to know what firm (or individual) the developer intends to hire as the property manager.

3.2.2. Retail

There are several different kinds of traditional retail properties, which are generally classified based on their size, location, and mix of tenants. Neighborhood retail developments are comprised of a small collection of basic-needs providers that serve a community, such as dry cleaners or grocery stores. Strip centers are collections of shops, service businesses, and restaurants in which the stores are traditionally arranged in a long strip with exterior doors facing along a pedestrian walk way. Malls are the largest form of retail development and may be either enclosed in a single, large building or feature external stores arranged along several pedestrian "streets." Many retail developments also include retail pad sites. *Pad sites*, or *outparcels*, are areas on the periphery of the development designed to accommodate a smaller, freestanding building with dedicated parking, as shown in Fig. 3.3. They are often located close to the main access road and have higher visibility than other stores in the same development. As a result of having a separate identity, higher visibility, and parking, the rental rate for pad sites is typically much higher than for other store space. The most common pad site users are restaurants, banks, or gas stations whose business models support higher rental costs. Although part of a larger development project, pad sites are often

FIGURE 3.3   Sample retail pad site. (*Image modified from Fairfax County GIS data.*)

located on individual parcels that have been intentionally subdivided and separately deeded from the rest of the development. This allows the pad site to be either ground leased or sold to a retail user, generating revenue for the developer and further value to the retailer.

Most larger retail developments will feature one or more *anchor tenants*, which are financially strong businesses that are popular enough to draw large numbers of customers to a shopping area. Although not all anchor tenants are of equal caliber, having at least one is important to the developer because anchor tenants (1) attract shoppers to the development, where they may then choose to spend additional money at other stores, and (2) ensure that at least one tenant in the retail development is able to pay rent in an economic downturn so that the developer (or future investor) has continued cash flow to cover, or at least offset, operating costs. Note that because of their importance to a retail development, anchor tenants often have a strong position when negotiating lease terms. Further, a developer usually cannot secure the necessary funding to begin a project until he/she has leases signed with significant anchor tenants to demonstrate the project's viability.

Anchor tenants are also usually destination retailers. *Destination retail* is a type of store that customers make a deliberate decision to patronize and will drive to largely regardless of its location.

In a neighborhood retail center, grocery stores are examples of both anchor and destination tenants. Households need groceries and, despite increasing rates of Internet shopping, consumers still prefer to buy groceries in stores.[3] Therefore, the majority of customers make the decision to physically go to a grocery store to shop. Certain types of modern destination retail are becoming increasingly more synonymous with *experiential retail*, in which the retailer has intentionally created an enhanced shopping experience to provide the customer with an experience that is also enjoyable and/or entertaining. This may include activities, dining, interactive features, social/community components, music, or other experiences. Experiential retail evolved from the need for physical stores to differentiate the shopping experience in order to compete with Internet sales of goods. In contrast to both destination and experiential retail, *convenience retail* describes stores that shoppers spontaneously decide to visit, usually once they are already in the store's vicinity. Site visibility, access, and traffic counts on adjacent streets are important factors for retail developers to consider when choosing a convenience retail site. Generally, better visibility and access with higher traffic counts are desirable features for any retail project; however, these are especially important for projects featuring convenience retail that attracts people largely on impulse.

The rental structure in many retail leases is different from most other product types. Two common kinds of retail leases are triple net leases and percentage rent leases. A *triple net lease* (abbreviated as NNN) is one in which the rent payment does not include (or is *net* of) the cost of three major expenses: insurance, taxes, and maintenance. These costs are paid directly by the tenant, as are utilities. The NNN structure tends to work best on freestanding, single-occupant buildings where there is no confusion about what portion of cost is attributable to which tenant. Pad site locations often feature triple net leases. When paying *percentage rent*, the retailer pays both a minimum monthly rent and also pays an additional rental payment equal to a specified percentage of gross sales over a pre-negotiated threshold amount, or *breakpoint*. Because developers sometimes participate in the financial success of retail tenants through percentage rent arrangements in this way, they have a vested interest in the profitability of the stores in their developments. Under either lease structure, retail tenants pay their proportionate share of charges for overall property maintenance known as *common area maintenance* (CAM).

Unlike many other product types, developers usually provide retail space to tenants in very basic condition. Space with no interior improvements except drywall, electrical wiring, plumbing, and HVAC systems in place is known in the industry as a "vanilla shell." Any changes or customization to this basic space are known as a tenant's "build-out." Retail tenants are expected to build-out the interior of their own spaces. This allows each retailer to use fixtures, interior layout, and other design elements to create a customized experience for its customers. This also allows restaurants to create their own kitchen space with such specialized equipment as may be necessary for their business. Generally, specialized retail equipment consists of three main elements: furniture, fixtures, and equipment (FF&E), although this term can apply to other product types as well. While the developer does not directly bear the cost of the build-out, retail tenants with long lease terms may negotiate an extended free rent period. During this time, the tenant has possession of the space but pays no rent to the landlord. This helps the tenant offset costs while the space is being built-out or necessary tenant licenses, such as a liquor license, are being obtained.

Figure 3.4 provides an example of a common site layout for a retail center on the Atherton Road site.

Considerations for Developers. Reconciling retailer requirements with restrictions in zoning ordinances can be problematic for developers. Examples of such challenges are abundant. A project located in an historic overlay district that requires architectural review and prohibits signs with more than two colors may preclude retail tenants whose corporate branding scheme requires colorful signage. Similarly, a potential retail tenant may require that all their store facades use a particular material or incorporate specific, brand-identifying features; however, the local jurisdiction may bar the use of the desired material or features. In both instances, the developer will lose the tenant unless he/she can either obtain a variance/special exception to the zoning or convince the tenant to modify its corporate requirements. In many cases, neither resolution may be possible.

Retail challenges are not tied exclusively to branding features. Space/use challenges also exist. For example, a grocery retailer may require a certain minimum-sized space for its grocery stores that exceeds the maximum allowable size for any building in a particular shopping center under applicable zoning ordinances. In such a case, the developer may need to decide if rezoning the property is feasible. It is also possible that a site simply cannot accommodate a retailer's unique space requirements, particularly if a particular user has a certain, designated footprint for all its facilities. If such a tenant preleases space and works with the developer during the early stages of project planning, it may be possible to adjust the site layout to accommodate the retailer's needs. However, if the project is already under construction or significantly preleased to other tenants, it may not be possible for the developer to alter the site.

Considerations for the Design Team. Site access and circulation must be carefully considered for retail projects. In the concept design phase, it is important to determine access to the site and coordinate with the local department of transportation (DOT). Access location and management

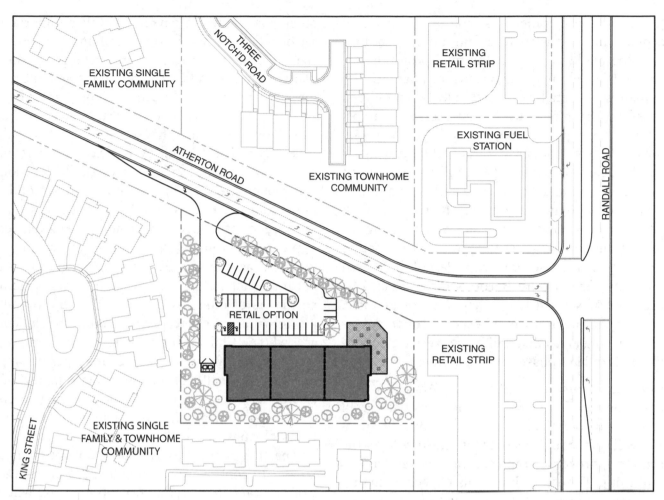

FIGURE 3.4   Atherton Road: sample strip center retail layout.

decisions are based on conditions pertaining to the road(s) adjacent to the site and on the number of trips expected to be generated by the finished development. Large retail centers will often require improvements to the adjacent roadway, such as the creation of turn lanes or installation of traffic signals, in order to accommodate the new development. These may not only be necessary for the success of the project but may also be transportation infrastructure improvements required for development approvals, as discussed in Chap. 2. Site restrictions, offsite improvements, and other transportation requirements can have a significant impact on the feasibility of a retail project.

In addition to access to and from the adjacent roadway, parking movements and access through the site are especially important considerations for retail developments. Poorly designed

sites can lead to internal traffic congestion and pedestrian-vehicle conflicts. Parking requirements have significant impact on retail site design and it is common for parking lot space to be greater than building space in terms of total lot coverage. Parking requirements are generally set by the local jurisdiction, but retail tenants often have their own, additional parking requirements based on customer profile. Further, tenant deliveries must be accommodated. The entire internal road network of the retail center may be shaped by the need to accommodate the access of especially large trucks to the tenant loading areas for deliveries. Emergency vehicle access and loading circulation must also be evaluated. Fire lanes are often required to be marked with signs and paint, and movement of fire trucks may need to be shown in design documents to prove access.

Phased development is common for retail centers and infrastructure and utility decisions must be made to support the final project build-out, not just a particular or current phase. Utility demands for retail centers should be considered based on the anticipated tenant uses. Restaurants often require grease traps and additional capacity for water, sewer, or gas. If parts of the retail center will be developed as pad sites, utility connections must be coordinated to the pad and for the tenant. Phased development also means that even though all or large portions of the site may be designed during the early stages of the project, construction may not occur for several years. Part of the phased planning may require portions of a site also be designed to serve a temporary use. For example, a new retail site may feature excess temporary surface parking areas that will be converted to building pads in later phases of development.

Considerations for Attorneys. Attorneys will help the developer review and negotiate retail leases with potential tenants. For example, an attorney will want a retail lease to give the developer the right to approve plans and contractors for a tenant's build-out. The lease should also specify whether or not the tenant will be required to remove any FF&E or other specialized improvements made to the shell space at the end of the lease term. With respect to liability, attorneys want to ensure that their developer-clients are protected and held harmless for the actions and activities of the retail tenants. Similarity, the lease will require tenants to maintain certain levels of liability insurance.

There are several important lease clauses that are often unique to retail projects, including percentage rents as previously discussed. Another such clause is a "noncompete clause," which prevents the developer from leasing future space to potential tenants operating like-kind or competing businesses to existing tenants. In some cases, noncompete clauses protect both the retailer and developer by ensuring that an existing tenant does not face undue competition and is able to stay in business (and pay rent). However, if a retail tenant insists on defining its business extremely generally, for example, as a "fast food restaurant" instead of a "chicken sandwich restaurant," such a clause can severely limit the developer's ability to identify other noncompeting tenants. Another important set of retail clauses are operating and "kick-out" clauses. *Operating clauses* require retailers, particularly anchor tenants, to remain open and operating. This ensures that the center is not blighted by a retail space that is occupied by a store that remains closed, or "dark." Note that so long as rent is paid, closed store space is still under lease to the retail tenant, which may decide to maintain the location and reopen the store in the future should conditions improve. Because the lease is still valid, in the absence of an operating clause, the developer is prevented from releasing "dark" or seemingly unused spaces to new, operating tenants. This can harm smaller tenants that rely on business from customers of others stores as well as the overall performance and perception of the development as a whole. While operating clauses can help prevent retailers from shuttering underperforming locations as a matter of corporate interest, they can also obligate retailers to continue to operate even when a store is legitimately losing money. The latter is obviously objectionable from the retailer's perspective; however, developers do not want a retailer to continue to occupy a space unless the store is open. To offset an operating clause while satisfying a developer's concerns, retail leases often also incorporate a kick-out clause. The *kick-out clause* allows a retailer to terminate its lease under certain conditions related to sales, economic hardship, or business performance. Thus, the kick-out clause is a mechanism that allows developers to keep only operating retail tenants in-place and reclaim and release space if a store chooses not to operate. Attorneys can help their developer-clients negotiate the specific terms of operating and kick-out clauses for the success of the project.

Other retail-related liability concerns for attorneys involve open space activities, such as an outdoor movie night at an experiential center. Such events can increase appeal for customers but also risk uncontrolled activities, such as self-catered alcohol consumption.

Considerations for Lenders. As always, lenders want assurance that the project will be successful and generate sufficient revenue to repay the loan. Therefore, before approving any financing for construction, lenders will often require the developer to achieve sufficient preleasing to demonstrate the project will be able to break-even on costs, including loan payments. Lenders will also scrutinize the leases for major tenants to determine if there are provisions that could negatively impact other tenants or the overall retail development, such as kick-out clauses. If the retail center is an existing project being acquired for redevelopment/repositioning, the lender will evaluate the reserves available for future capital improvements and also examine the leases to determine which leases (if any) will be ending in the near future and, therefore, what tenants may need to be replaced. In order to qualify for financing, the developer will need to demonstrate that the center generates enough cash flow to cover loss of rent, brokerage leasing commissions, and other costs associated with new leases.

3.2.3. Office

Office developments may be comprised of a single building, a multi-building office park, or corporate campus. Office properties are often described as either "suburban" or "urban;" they can be described by their height: low-rise, mid-rise, and high-rise. However, they are more often classified based on a combination of property age, quality, and level of amenities using the terms "Trophy Class," "Class A," "Class B," and "Class C." Although these asset classes can apply to other product types, they are most often discussed in the context of office properties. Although the Building Owners and Managers Association (BOMA) International offers some guidance, there are no firm definitions for each class and subjective standards are different across different regions and in suburban versus urban areas. Generally, however, the following guidelines can be used:

Trophy

- New buildings constructed within last 10 years.

- Incorporate the latest technology and are constructed with top quality materials, high-end finishes, and building materials.

- Feature extensive amenities such as a conference center, fitness center, roof top deck, and concierge.

- Are often transit orientated and developed on sites near metro locations in urban areas.

- Command premium rental rates.

- Tenants tend to be large corporations who can afford premium rents and may seek properties that fit their corporate image and the expectations of their clients.

Class A

- Newer buildings constructed within the last 10 to 20 years.

- Include some technology, are often energy efficient, and feature modern design.

- Numerous and desirable amenities.

- Typically have high-end finishes and features.

- Often well located.

- Feature a lobby reception/security desk.

- Command high-end market rent.

Class B

- Buildings are generally older (20+ years).

- Feature limited amenities and basic finishes.

- Rents are moderate and reflective of the lack of amenities, although there may be high demand due to a good location or affordable quality.

- Legacy tenants may still be in place on original, long-term leases from the building's delivery, while new tenants are often cost-conscious firms seeking the affordability of Class B buildings.

- Note that it is possible for a newly constructed building to be delivered as a Class B property if it has limited amenities and modest finishes, particularly in suburban markets; however, most new development, especially in urban areas, seeks to be Class A.

- A developer may choose to renovate a Class B building, but may also make a deliberate investment decision to maintain the building as Class B in order to capture market demand for affordable office space.

Class C

- Buildings are old and generally approaching the end of their useful lives.

- Limited or no amenities.

- Noncompliant with current codes, for example, ADA.

- May suffer from lack of maintenance and investment.

- Older building systems and construction methods are usually inefficient.

- Often comprised of many small suite spaces.

- Offered at a discounted rent as a result of age and lack of amenities.

- Tenants may be nonprofits or nontraditional office users.

- Well-located Class C buildings often represent repositioning/redevelopment opportunities for developers.

Unlike retail tenants, office tenants are often provided with a build-out budget, known as a *tenant improvement allowance*, which is incorporated into the value of the lease. The tenant may use an independent interior design firm or work with the developer's design team to create a *test fit* that describes how the office suite will be laid out to accommodate cubicles, individual offices, break rooms, conference rooms, file rooms, and other spaces to meet the tenant's needs. The test fit is modified as needed and converted into a *space plan*, which is then used to guide the work for the tenant's build-out, or *tenant improvements* (TI), which are largely paid for by the TI allowance. If the cost of the build-out exceeds the TI allowance, the office tenant will have to contribute funds to cover the difference or request that the developer include the additional costs into the lease terms by increasing the rental rate. Most tenants typically require the developer to build-out the suite space on their behalf.

The developer must budget for TI costs not only for initial tenant occupancy, but for each time an individual tenant's lease expires (along with brokerage commissions and other costs associated with releasing space). Developers attempt to amortize the cost of each tenant's TI over the term of their particular lease; thus, the amount of TI allowance offered to each tenant depends on the length of lease term, amount of rent, and strength of the tenant. Developers willing to offer large TI budgets may be able to use this as an incentive to attract desirable tenants to their building. In competitive markets, tenants renewing leases in their existing space are often given a modest TI allowance to "freshen" up the space and make improvements. The size of a lease renewal TI allowance depends on market conditions and the degree to which the developer wants to incentivize the tenant to remain in the building.

Office buildings are tenanted by various types of companies ranging from traditional office users, such as accountants and law firms, to more specialized and creative users, like technology companies and advertising firms. Medical providers are another specialized form of office tenant with unique needs. For example, medical space generally requires additional plumbing requirements, floor load capacities to support especially large or heavy equipment, and extra parking because of the number of daily patient visits. As a result, medical uses cannot easily be accommodated in typical office buildings and often congregate in dedicated medical office buildings built by specialty developers. The segregation of medical tenants is often further encouraged by zoning regulations that restrict certain uses. When medical (or other) space is particularly expensive, developers may be willing to spend additional money beyond the budgeted TI allowance by amortizing the additional cost into the tenant's rent and increasing the rental rate accordingly.

As with other product types, office tenants are charged a lease rental rate based on the *rentable square footage* of their suite space. An office tenant's total rentable square footage is comprised of two elements: (1) the actual square footage used by the tenant, or *usable square footage*, and (2) the tenant's proportionate share of the building common areas, such as hallways, lobbies, elevators, and amenities. The square footage of the common area is known as the building's *core factor* and can vary across different development projects. Most office leases follow a "full service" rental structure. A *full service lease* is one in which the tenant's monthly rent includes all expenses attributable to their total rentable square footage. The developer uses a portion of the rent collected to pay operating expenses, including building taxes, utilities, and maintenance. Although the cost of these operating expenses is fully included in the first year's rent, also known as the *base year*, as operating costs increase tenants must pay their proportionate share of increases over base year costs. In addition, landlords will often also escalate the rent each year by a small percentage, to help cover inflation and other unknown outside costs. Because the developer collects a fixed rental amount, he/she is usually highly motivated to control and limit costs while still maintaining the building.

When market demand is weak, office developers may have to offer concessions to attract tenants. Other than increasing the TI allowance, developers may offer a rent-free period. This can range from a single month to many months. However, "free rent" is only free to the tenant; the developer must still pay all the same operating costs for the rent-free period with no rental income to cover costs. The impact of such an incentive is evaluated by calculating the *effective rent*, or annualized rate per square foot the developer actually collects from the tenant.

Figure 3.5 provides an example of a potential site layout for an office development on the Atherton Road site.

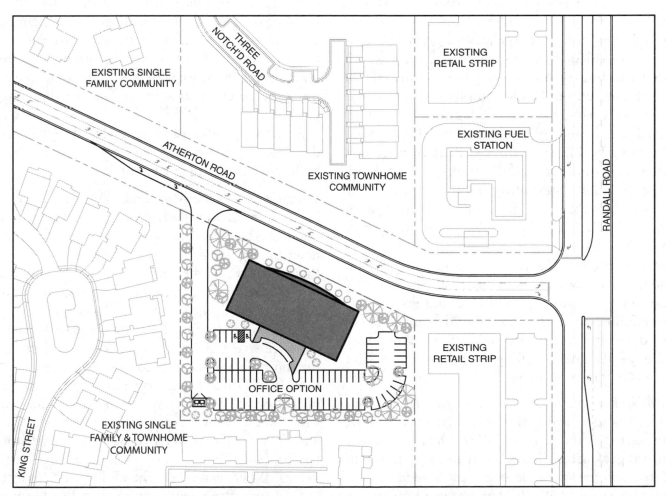

FIGURE 3.5   Atherton Road: sample office layout.

Considerations for Developers. When preleasing new office product, developers will often try to lease out entire floors without "breaking" the space to accommodate multiple smaller tenants. When possible, the developer may rent the entire building to a single, or very few, stable and extremely high credit tenants, such as Fortune 100 firms. However, if a single-building tenant leaves or merges with another company, the developer may become the owner of a vacant building, which jeopardizes project returns if a replacement tenant(s) is difficult to find. Therefore, from a strategic standpoint, some developers prefer to have a combination of large tenants renting an entire floor (or floors) and small tenants that rent different suites on other divided floors. This mix of tenants can reduce the overall risk of loss of rental revenue if one tenant leaves or defaults.

Many jurisdictions are now requiring that new office construction qualify for certification in Leadership in Energy and Environmental Design (LEED) by the U.S. Green Building Council (a nongovernmental organization). Green Incentives provided by the jurisdictions, such as FAR density bonuses, may or may not compensate for the higher cost of LEED compliance. Tenants are also increasingly becoming familiar with LEED certification and will sometimes require that their building be LEED certified. LEED certifications increase project costs but can often help a developer position a new office building to command rental rates that are higher than market averages. Accordingly, office developers often incorporate LEED requirements into the design and budget of new projects. In addition to leasing benefits, some real estate investment trusts (REITs) only consider investing in LEED buildings in their portfolios, so a merchant builder may choose to develop a LEED certified office building with the specific goal of selling the property to a REIT after the property stabilizes.

Considerations for the Design Team. Site access and coordination for office projects are impacted by the density of the project, anticipated use at peak access hours (8 am and 5 pm), and the large volume of single occupant vehicles generally consistent with the employees of office tenants. Office developers often have a specific type of office tenant in mind for their projects, which can also have some impact to the site layout. For example, a medical office building may require a greater than average number of parking spaces while a technology-driven tenant may require additional communication capacity and redundant sources of electricity. Utilities are usually provided with a single connection point for office developments; however, larger or high-rise buildings may require additional fire protection and water supply. Stormwater management is usually a common system for the office development. In some cases, the stormwater management can be a site amenity such as a large central pond within an office complex.

Office parking requirements are generally lower than retail parking requirements but can still be difficult to accommodate, depending on the site. In urban areas where land is expensive or site space is limited, parking may be accommodated either in a subsurface structure below the office building or in an on-site multistory parking garage adjacent to the building. Both parking solutions are more expensive than surface parking and the design team will work with the developer to determine the best way to accommodate the building program.

Structural loads and other vertical design coordination issues are of greater concern for office buildings, which are often multistory.

Considerations for Attorneys. While all leases have common provisions, office leases differ from retail leases in certain key areas. For example, determining a retail tenant's floor area is different than determining the rentable area of office space; office leases may specify the measurement standard to be used, such as that produced by BOMA. As previously discussed, the total square footage of office space includes a core factor to account for common areas such as lobbies, restrooms, and corridors. Office leases typically permit general office uses and may have less narrowly defined uses than a retail lease. Additionally, office tenants may not be as likely to want to restrict the landlord's ability to lease to competing businesses. Whereas a retail tenant's location may be critical, an office tenant may be more flexible and may not resist the inclusion of a relocation clause in the lease as strongly as a retail tenant. This is not always the case, however, and an office tenant leasing an entire top floor or other premium space may be opposed to the possibility of future relocation.

Office landlords often do not require "continuous operations" for tenants because, unlike in the retail setting, an office property is less affected by a tenant "going dark" while continuing to pay rent. Given that tenant mix may be less important to the success of office buildings than retail developments, office tenants are often able to negotiate more generous assignment and subletting provisions, so long as they can assure the landlord of the creditworthiness of any assignee or sublessee. In the retail setting, tenants often build out their own space according to their "corporate architecture"; however, in office buildings, landlords often control the tenant build-out. In such cases, the tenant's obligation to begin paying rent is usually tied to the landlord delivering the premises to the tenant with the landlord's work "substantially complete," which is a term that should be defined in the lease.

Considerations for Lenders. As with other product types, lenders will often require sufficient preleasing to demonstrate project viability. While developers may want to find the largest possible tenants for their buildings, lenders may take a different view. If a project's financial viability is tied exclusively to the rent paid by a single tenant, or few tenants, the project is immediately jeopardized if the tenant(s) encounters operational trouble, files for bankruptcy, or is otherwise lost. Such concern on the part of lenders is not without merit. Federal agencies, which often lease large office spaces or entire buildings, have a unilateral government right to leave a lease. This same right can also apply to state government tenants. Further, recent history provides plenty of examples of large, private corporations previously assumed to be quality-credit tenants that effectively "disappeared" overnight, such as Enron, Arthur Anderson, and Lehman Brothers, all in the 2000s. Finally, while kick-out provisions are less common in U.S. office leases, they do exist and lenders will want to review any such clauses, especially for large or entire-building tenants.

3.2.4. Industrial

In real estate, the term "industrial" includes a wide range of uses including warehouses, distribution centers, manufacturing plants, laboratory space, and even limited office space. Beyond these more "traditional" industrial uses, a new specialized use has recently been added to the industrial category: data centers. These are typically large buildings with stringent requirements for redundant and reliable power supply, robust Internet connectivity, and access to water for cooling. Industrial property types can be further subclassified based on the type of activities that occur on the site. For example, a plant in which the manufacturing process uses large, energy-dependent equipment, processes toxic chemicals or other raw materials, and produces significant waste, may be referred to as "heavy industrial" whereas a facility where existing components are assembled into a final product might be considered "light industrial". Warehouses are also often subclassified based on their size and specific use. For example, a specialty warehouse that requires extensive refrigeration, usually for food or other perishable products, is often called "cold storage". "Fulfillment centers" or "distribution warehouses" are short-term storage facilities where products arrive, are sorted, and promptly shipped out to consumers. A building that is designed to be flexible enough to include significant office operations in combination with a range of supporting activates, usually storage space or laboratory space for research and development, is referred to as a "flex" building.

Several features are common across different industrial uses. For example, proximity to major ports, interstates, and/or delivery arteries is important for most industrial tenants. Also, all types of industrial uses will generally require site access and loading docks for large trucks, although this need may be less critical for flex tenants using primarily local delivery/work fleets that use drive-in doors. Conversely, some industrial uses have unique requirements. For example, heavy manufacturing facilities require substantial power supply and may also require greater than average water usage. Heavy manufacturing users often need especially tall ceilings to accommodate indoor cranes, lifts, or other equipment. Distribution centers may use special inventory storage and retrieval equipment that requires particular space for internal drive passages and *clear-heights*, or the distance between the floor and lowest roof structure.

From a planning perspective, industrial properties are generally considered undesirable neighboring uses and relegated by zoning ordinances to congregate in specially designated industrial parks. This allows noise, fumes, contamination, and other negative externalities from industrial operations to be collocated in areas away from the general population.

Considerations for Developers. While some types of highly specialized industrial space, such as laboratories, can be costly, the construction of many industrial buildings, particularly basic warehouse, is less expensive than many other product types. There are several reasons for this including: significantly lower use of window glass; more open interior space and less build-out; single-story design; less priority on "representational space" such as lobbies or atriums; fewer amenities; and less expensive materials for facades and fixtures. This also means that construction can often be completed more quickly for industrial projects than for many other product types. However, industrial rental rates tend to be lower than office or retail rates, making project viability equally important to consider as with any other product type. Zoning, environmental, and other regulatory hurdles apply equally to industrial projects, although industrial developers need to pay particular attention to regulations pertaining to waste disposal, noise, and other negative externalities as these may impact how future tenants can use the property.

Considerations for the Design Team. Design for basic industrial buildings is often relatively simple compared to other product types. Unique design considerations involve accommodating extensive heavy truck traffic, including cross-loading docks as necessary, and potentially substantial utility demands for manufacturing properties. One important design feature of industrial buildings is the ratio of the depth of the building to its length. Certain types of industrial tenants require large, deep open spaces. However, smaller flex tenants cannot "fit" properly in a deep, cross-docked facility.

Considerations for Attorneys. With industrial development projects, attorneys are concerned not only about existing environmental contamination and liability when the developer purchases land, particularly if it is a brownfield industrial site, but also with respect to future contamination from tenant operations. To mitigate damage by tenants, industrial leases can require that tenants purchase both general liability and environmental liability policies. Further, attorneys may recommend that their developer-clients purchase special environmental impairment liability insurance policies.

Concerns for Lenders. Given the nature of activities that occur in industrial properties, lenders are particularly sensitive to environmental concerns. For example, heavy manufacturing or even extensive automotive uses could create additional risks to a bank if forced to foreclose and take the property as collateral. Under such circumstances, the bank would then assume the environmental remediation risk upon taking ownership of the property. Environmental damage to the site that requires remediation limits the property value and potentially carries liability risks.

3.2.5. Mixed-Use

The term "mixed-use" has evolved over the decades with respect to real estate. Before regulation of land use through zoning ordinances became prevalent, simple examples of mixed-use were relatively common: a storeowner might live in a residential unit above his/her shop in a small two- or three-story building. However, with the rise of strict Euclidean zoning regimes, sites that contained more than one product type became harder to develop. As zoning has evolved to more readily facilitate modern lifestyles, today's comprehensive plans now permit large developments to combine a variety of uses including hotel, retail, residential, and office. These are generally referred to as mixed-use developments. However, many such projects are limited to a collection of single-product buildings. For example, a hotel might be included in the planning of a large destination entertainment-retail project but be physically separated from other uses, perhaps occupying its own pad site. Similarly, a large "town center" style project might include hotel, office and residential towers, as well as retail and entertainment venues. Yet, with the exception of including first-floor retail space beneath office or residential buildings, the different product types are usually accommodated in separate buildings. Growth of true *vertically mixed-use* projects, in which multiple uses are stacked in the same building is now on the rise, particularly in urban markets.

Considerations for Developers. Mixed-use developments are becoming increasingly popular as demand for walkable, work-live communities continues to grow. However, mixed-use projects tend to be larger, more expensive, and more complicated than single-use projects. Due to their complexity, it can also take longer to satisfy municipal examination and obtain approvals for mixed-use projects, even when such growth is encouraged by comprehensive planning exercises. Large mixed-use projects may also require the additional cost of special "place making" consultants to help plan a deliberate mix of uses, architecture variations, amenities, and other components to create an appealing overall environment. Thus, capacity and funding limitations preclude many developers from undertaking mixed-use developments.

Vertical mixed-use projects present additional design and operational challenges. For example, while there may be market demand in the same area for both a hotel and residential units, combining the two uses in a single building can produce undesirable results. If a vertically mixed-use building features hotel rooms on floors one to five and residential units on floors six to 10 with a shared lobby and elevator system, residents and guests will be intermingled on the first floor. Residents returning home after work may not want to wade through a lobby full of suitcases and travelers. For residents with young children, there could also be safety concerns over the constant stream of strangers in the building. Potential incongruences increase if the mix includes a combination of office use, as office professionals may prefer not to share lobby space with sick residents staying home from work or taking their dogs out for a walk in pajamas on their days off. Of course, there are measures developers can take to alleviate the conflicts of

vertically mixed-use buildings, such as restricting floor access or designing separate entrances, lobbies, and elevator spaces for different uses. However, any vertical mixed-use solution will be more costly and require more planning and future management than simply developing a single-use project.

Considerations for the Design Team. Design and engineering concerns for mixed-use projects generally reflect the considerations of their individual product components (residential, office, retail, etc.). In some cases, overall parking requirements might be reduced because of the timing variation in parking needs of different uses; for example, the same parking spaces can be used by office tenants during the day when residents are gone and vice versa. Many mixed-use developments are transit oriented, which means reduced vehicle parking capacity may be justified. Further, alternative modes of transportation should be considered and facilitated such as bike and pedestrian routes. Similarly, internal road networks on mixed-use sites generally have few and narrow lanes, with on-street parking and large sidewalks. This design encourages low-speed driving and supports the pedestrian or nonmotorized transportation.

One of the greatest challenges with mixed-use developments is the large scale of the project in terms of site size, time, components, and the development team itself. Indeed, because of their scale, mixed-use projects usually occur in phases. However, changes in market conditions over time may lead to the need for design changes between the initial and later phases of construction. Additional coordination by the design/engineering team is required throughout the extended design and construction periods.

Considerations for Attorneys. Mixed-use projects are as complex from a legal perspective as from a development perspective. The entitlement process may be more extensive than for a smaller, single-use project; however, it may also provide greater opportunity if local planning documents and zoning ordinances allow more flexibility for mixed-use projects. After the entitlement stage, such projects generally involve have more "moving parts" than a development for a single use due to the inherent complexity of mixed-use development. For example, there may be multiple contractors for different components of design and construction, several different kinds of tenant leases or sales contracts, and the project may also involve more complex partnership agreements for the developer. Further, some uses may interact with others in ways that require special treatment. The developer's attorney will help coordinate the array of contracting and legal actions involved.

Concerns for Lenders. When underwriting a large, horizontal mixed-use project, lenders may seek to provide financing in phases. This risk mitigation strategy ensures the first project phase is successful without stretching demand by building-out future phases or draining the developer's resources (available cash, staff, labor, time). Assuming the first phase(s) generates sufficient cash flow and the developer can demonstrate continued demand, the lender will continue to finance additional phases. However, vertically mixed-use projects cannot easily be built in phases and lenders must engage with the entire project concept at once. This often requires specialized lenders who understand different product types and are comfortable evaluating complex projects.

3.2.6. Specialty Projects

There are many, many uses for buildings other than those described in this chapter. The list of specialty product types is extensive, including (but certainly not limited to) unique uses such as the following.

Medical

- Healthcare (generally)
- Memory care
- Assisted living
- Hospice care
- Hospitals

Hospitality/leisure

- Hotel
- Bed and breakfast
- Resorts (ski, beach, etc.)
- Spas

Entertainment

- Sports stadiums and sports fields
- Racetracks
- Specialized sports/gym/fitness centers
- Aquariums
- Museums
- Theme parks
- Movie theaters
- Golf courses and club houses
- Shooting ranges

Infrastructure

- Airports
- Seaports
- Schools

Animal/livestock

- Veterinary offices and kennels
- Animal rescue facilities
- Stables
- Livestock and meat processing facilities

Other

- Refineries
- Consumer self-storage centers
- Religious places of worship

Some specialty uses fit, albeit often imperfectly, into one or more of the "traditional" use categories, such as retail or office. However, just as often, specialty-use developments confound rigid zoning regimes and often require special permissions, zoning or otherwise. Many large specialty projects, like hospitals or stadiums, are extremely expensive and can cost upwards of $1 billion for a single project, which may necessitate a unique lending and investment structure. Before undertaking any specialty project, developers must objectively evaluate whether or not the necessary specialized expertise exists both on the part of the developer and the different members of the development team.

3.3. KEY CONSIDERATIONS FOR DEFINING THE PROJECT

Developers will often specialize in one (or a combination of) the wide range of development projects and product types described thus far in this chapter. The nature of the type of proposed project can affect the size, type, and location of sites that may be appropriate to pursue. However, these are not the only considerations developers must be aware of when seeking opportunities. The rest of this chapter will discuss developer capacity, the real estate market lifecycle, and early project planning constraints and their implications for identifying development opportunities.

3.3.1. Evaluation of Capacity and Opportunity

Every developer needs to evaluate his/her capacity in order to structure the search for potential opportunities. This is true regardless of whether the developer is following an investment committee mandate to place an equity raise, entrepreneurially looking for a site and/or tenant, already owns a site, or is working through a prescribed public sector acquisition planning process. Lack of an accurate and objective understanding of the developer's own capacity can easily lead to disastrous results in instances where the developer lacks the competency or resources to complete a project after it has been started.

Scope, Budget, and Developer Specialization. Not all available sites represent actual opportunities to a developer. Each developer must assess his/her own capacity with respect to scope, budget, and area of expertise. Failure on the developer's part to accurately understand these limitations can lead to financially risky projects with a higher rate of failure. The following examples will manipulate the sample site introduced in Chap. 2, Atherton Road (Site A), to demonstrate several different scenarios.

Example 1
The developer must make a preliminary estimation of project costs in order to determine what scope and budget is appropriate for his/her individual capacity. To demonstrate this exercise, we will consider the cost difference between two project opportunities. Imagine the Atherton Road site (Site A) is a 1-acre site zoned for low-rise/low-density office development as shown in Fig. 3.6. The second site is located on Baldwin Court (Site B) and is also zoned for office use, but is a 5-acre site. The current zoning for both sites allows a 0.5 FAR by right density development.

Recall from Chap. 2 that FAR refers to *floor area ratio*, a bulk zoning restriction for commercial properties that dictates how much building space can be accommodated on a particular site. FAR is the coverage ratio of total building square feet (sf) to total site area. Based on the FAR of 0.5, the developer can build an office building with a gross area of 21,780 sf on Site A and 108,900 sf on Site B, calculated as follows:

Site A:

$$1 \text{ acre} = 43{,}560 \text{ sf}$$

$$43{,}560 \text{ sf lot size} \times 0.5 \text{ FAR} = 21{,}780 \text{ sf building}$$

Site B:

$$5 \text{ acres} = 217{,}800 \text{ sf}$$

$$217{,}800 \text{ sf lot size} \times 0.5 \text{ FAR} = 108{,}900 \text{ sf building}$$

Once the developer knows the approximate size of each building, he/she can make preliminary assumptions to arrive at an estimate of project cost. This is essential for determining what scope of project the developer should pursue. The costs involved with any development project are generally broken down into three categories: hard costs, soft costs, and land costs. *Hard costs* are tangible items directly related to the construction of the building. Hard costs also include associated labor costs of construction and contingencies for hard costs overruns. *Soft costs* are necessary but indirect costs required to complete the project. Figure 3.7 provides some

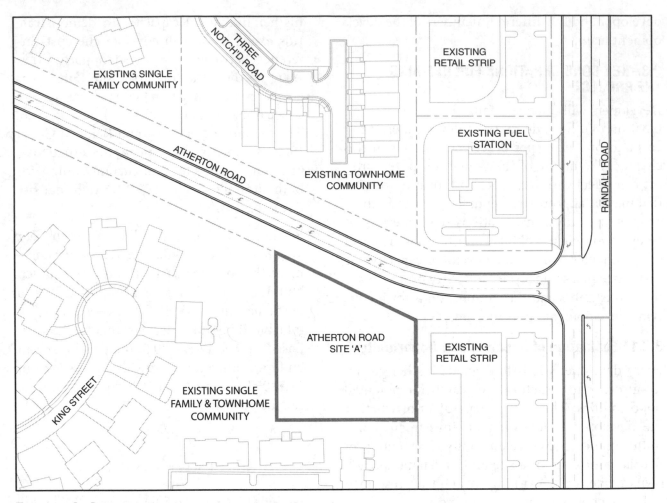

FIGURE 3.6 Atherton Road: 1-acre site with office zoning.

examples of typical hard and soft costs. *Land cost* refers to the cost to acquire land upon which the development project will be built. Note that the purpose of this example is to highlight the magnitude of the cost difference between buildings of different sizes, so land cost will be excluded.

In most cases, it will not be possible to include exact numbers for each line item at this early stage in the development process because many decisions have not yet been made about project components and building design. Therefore, the developer will use educated estimates of these costs rather

Hard Costs	Soft Costs
Building envelope (window glass, façade, roofing, etc)	Architecture/design fees
Building foundations	Attorney/legal fees
Building interior finishes (carpet, paint, doors, etc)	Engineering fees
Building materials (concrete, brick, etc)	Insurance costs
Building systems (electrical, plumbing, HVAC, etc)	Loan interest (or "carry costs")
Construction labor costs	Marketing costs
Landscaping materials and labor costs	Permit and inspection fees
Site work (excavation, grading, fill, etc)	Surveys and studies
Utility work	Taxes

FIGURE 3.7 Sample hard and soft costs.

than take the time (and incur cost) to price each line item. Typically, soft costs account for approximately 10 to 15 percent of total hard costs, although this can vary depending on product type, location, and whether or not the site must be rezoned. Hard costs can vary tremendously by product type as well as by location. For example, the average office construction hard cost per square foot in San Francisco is far greater than the average cost in Indianapolis.[4] Note that at this early stage it is a best practice to include an estimate for contingencies and potential costs overruns, usually between 5 and 10 percent of total costs. This contingency can be refined later and may be only applied as a percentage of hard costs.

For this example, we will assume hard costs are estimated to be $100 per square foot (psf) and soft costs are equal to 10 percent of total hard costs. We will also include a contingency allowance equal to 5 percent of the total construction cost. This produces an estimated project cost for each site, excluding the cost to acquire the land, as shown in the calculations in Fig. 3.8.

This rough cost estimating exercise allows developers to make a preliminary determination about their capacity and what project scope may be appropriate to pursue. Clearly, a development project on Site B requires a larger budget and greater developer capacity. Although both properties may be available for purchase, Site B in this scenario may not represent a realistic opportunity for a new or small entrepreneurial developer. Conversely, both sites may be too small for a national developer with substantial funds to invest. Note that once land costs are included, the total overall cost will obviously be greater.

This is a simplified example to help clarify why a developer may have chosen to pursue a smaller project/site that may seem less desirable from a technical perspective due to site conditions or layout challenges. Increasing project size, especially in areas with high land costs, can have a significant impact on the developer's budget. Of course, cost is not the only factor developers consider; other financial considerations will be discussed in future sections.

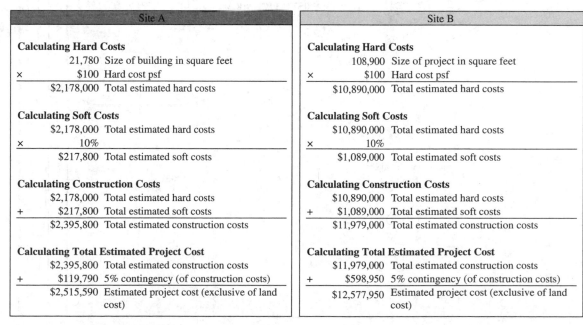

FIGURE 3.8 Hard and soft cost sample calculations.

Example 2

Although a site zoned for office use was used in the previous example, the same budget and scope considerations apply to residential developments. Now imagine the Atherton Road site (Site A) is actually 200 acres with medium density R-6 zoning, which permits up to six residential units per acre, as shown in Fig. 3.9. The site can accommodate 1,200 homes, making the scope of the project extremely large. A project of this size and type is both complex and costly, which likely means it is again beyond the capacity of new or small developers who do not have access to sufficient funds or the past performance to justify the endeavor.

Note that developer capacity is not the only reason a site may be infeasible. A mismatch between a site and a developer's area of specialization can also be a relevant factor. For example, a large residential developer might be able to handle the cost and scope of the 200-acre project described in this example. However, even if the R-6 zoning allows multifamily units, the Atherton Road site does not represent an opportunity for a multifamily developer who specializes in building high-rise projects in urban markets. Such a developer may not have the market expertise to accurately assess pricing, amenities, or demand for apartments that are not transit-accessible and do not have the surrounding community amenities of a dense urban environment. Thus, the location of Site A in this example is not a match for an urban multifamily developer's area of specialization.

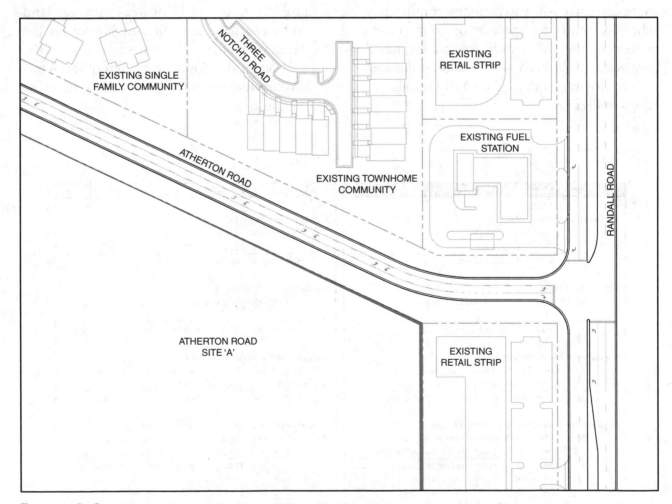

FIGURE 3.9   Atherton Road: 200-acre site with residential zoning.

Similarly, a sophisticated office developer working with a client to create a corporate headquarters campus might be looking for large sites such as the one in this example. However, such a developer may not have the time or expertise necessary to pursue a rezoning of the site to allow an office project. While the size and location may be acceptable, again, the site is not a match for the developer's area of specialization.

For large residential developers and homebuilders specializing in creating suburban communities, Site A in this example *may* represent an opportunity, depending on the cost considerations, market area demand for housing, the state of site entitlements, and many other factors.

Example 3

The 200-acre Atherton Road site is an extreme example to highlight a concept; however, not all scenarios are as clear. For example, imagine the original 2-acre Atherton Road site, as described in Chap. 2, has topography challenges that are partially hidden by loose fill because the site was historically used as a dumping location for construction debris from the neighboring townhome development years ago. The scope challenge is no longer one related to site size or location, but rather one of site conditions. The developer will face extra costs associated with removing the fill and designing to accommodate a potentially extreme grade, which could require incorporating expensive retaining walls, embracing an inefficient building footprint, or extensive site work.

Seasoned developers almost always have examples of projects they wish they had not pursued. Yet it is easy for a developer to believe he/she can take on a more complex project than may be realistic. This can occur any time the developer pursues a project in an unfamiliar location or market, misjudges the complexity and scope of a project, is overly eager to satisfy a client's demands, or underestimates the time requirements and consequences of risks involved. Any

of these scenarios can quickly lead to escalating project costs and timelines, both of which threaten a project's overall financial viability. In a worst-case scenario, a developer's insistence on pursuing an infeasible project can lead to contractual defaults, loss of funding, and the inability to pay the development team to move forward with the project. A good developer needs to establish an honest and objective review of his/her capabilities as compared to the scope and budget of a potential project.

Preliminary Zoning Considerations.

At the early stages of a development project, the developer may or may not be focused on zoning considerations, depending on his/her areas of expertise. For example, a developer familiar with the rezoning process may prioritize finding a well-located site at a discounted price and then seek a rezoning if the ideal site does not permit the desired project. Conversely, an entrepreneurial developer may simply look for a desirable site and then define a project based on whatever use is permitted. In either of these two cases, a site's designated zoning may not be a primary concern for the developer.

However, a specialist developer may only want to build a certain type of project and, therefore, may only be interested in sites where his/her product type of choice is already permitted by right. In this case, the developer will need to look at local zoning ordinances for definitions of various use types to determine whether a particular product type is permitted in certain zoning districts. Equally, a merchant builder that wants to build and sell as quickly as possible may be unwilling to dedicate the time and financial resources necessary to pursue a rezoning and will only consider sites with the desired zoning already in place.

Example 1

In additional to dictating property type, zoning can also impact project budget and viability through density. Therefore, even at the early stages of the process, it behooves developers to be aware of the different zoning designation in their markets. To highlight this, we return to the office development scenario in which the developer can purchase either the Atherton Road site (Site A) or the Baldwin Court site (Site B). However, for this example, we now assume that both sites are the same size: five acres. In this scenario, Site A is zoned for lower density development and permits only a 0.4 FAR. Site B zoning allows the same 0.5 FAR as in the previous example. By right, this allows the developer to construct buildings of the following sizes on the respective sites:

Site A:

$$5 \text{ acres} = 217,800 \text{ sf}$$

$$217,800 \text{ sf lot size} \times 0.4 \text{ FAR} = 87,120 \text{ sf building}$$

Site B:

$$5 \text{ acres} = 217,800 \text{ sf}$$

$$217,800 \text{ sf lot size} \times 0.5 \text{ FAR} = 108,900 \text{ sf building}$$

Using the same cost estimating exercise from the previous section, again excluding the cost to acquire the land, we can calculate the total project cost for each site as shown in Fig. 3.10.

The (approximately) $2.5 million difference in cost between these two projects is significant for some developers. For larger sites and/or higher allowable FARs, these cost differences would be magnified. Note that a smaller project, while less costly, also yields less rentable space and, therefore, generates less rental revenue (market conditions being equal). During more advanced stages of the project, the developer will need to create a more complete financial analysis taking into account market rates, land cost, and other factors. However, even at the early stages of the process, it behooves developers to be aware of the different zoning designation in their markets, as well as the costs and risks involved in rezoning, including the possibility that the jurisdiction's governing body will not approve the rezoning, whether due to flaws in the proposal, citizen opposition, or political pressures.

It bears mentioning that public sector developers are less likely to be concerned about zoning in the preliminary stage (or any of the following stages) because most zones already allow for public sector uses. Further, public sector development projects are not profit oriented and, thus, are not sensitive to density implications in the same way as private sector projects.

Site A		
Calculating Hard Costs		
	87,120	Size of project in square feet
×	$100	Hard cost psf
	$8,712,000	Total estimated hard costs
Calculating Soft Costs		
	$8,712,000	Total estimated hard costs
×	10%	
	$871,200	Total estimated soft costs
Calculating Construction Costs		
	$8,712,000	Total estimated hard costs
+	$871,200	Total estimated soft costs
	$9,583,200	Total estimated construction costs
Calculating Estimated Project Cost		
	$9,583,200	Total estimated construction costs
+	$479,160	5% contingency (of construction costs)
	$10,062,360	Estimated project cost (exclusive of land)

Site B		
Calculating Hard Costs		
	108,900	Size of project in square feet
×	$100	Hard cost psf
	$10,890,000	Total estimated hard costs
Calculating Soft Costs		
	$10,890,000	Total estimated hard costs
×	10%	
	$1,089,000	Total estimated soft costs
Calculating Construction Costs		
	$10,890,000	Total estimated hard costs
+	$1,089,000	Total estimated soft costs
	$11,979,000	Total estimated construction costs
Calculating Estimated Project Cost		
	$11,979,000	Total estimated construction costs
+	$598,950	5% contingency (of construction costs)
	$12,577,950	Estimated project cost (exclusive of land)

FIGURE 3.10 Hard and soft cost sample calculations for 5-acre sites.

Availability of Debt and Equity.   Once a developer has decided on a project scope and determined an approximate budget, he/she can begin to evaluate the availability of funding. This is a critically important consideration because in order for any development project to occur, funding must be available. The most desirable site in the world represents no opportunity at all if the developer does not have the financial resources available to purchase the land, pay for design and construction, and deliver and operate the finished building. It is the developer's sole responsibility to identify and secure the necessary sources of funding. Thus, at the early stages of a project, developers must establish a realistic idea of how much money they can source. While developers often make at least some financial contribution to their projects, the vast majority of funding comes from debt and equity sources.

As explained in Chap. 1, lenders are entities that provide a source of cash in the form of a loan, also known as *debt financing,* for real estate investment and development projects. All lenders, regardless of type, earn fees by collecting interest on the amount of the money they loan. Equity partners are private investors that provide money to developers in exchange for receiving preferred returns, which are higher than returns paid on standard debt financing. Equity investors can command this higher return because, unlike banks, these investors do not necessarily have any claims on the property title and, thus, take a higher risk of not recapturing their investment if the project fails.

Typically, there are two types of loans developers must obtain in order to bring a project to fruition: construction loans and permanent loans. *Construction loans* are typically interest-only loans that function similarly to a line of credit wherein the developer draws funds on an as-needed basis to cover the ongoing cost of site work and construction. Construction loans are usually paid off by the placement of the permanent loan, although some construction loans include terms that allow them to convert to permanent loans. *Permanent loans* are placed when the finished building reaches *stabilization,* or the point at which cash flow from rental income is sufficient to support operations and generate positive net revenues. The developer's construction loan and permanent loan can come from the same lender or from different funding sources. Because permanent loans remain in-place during the building's operating lifetime, the terms and availability of permanent loans are often of greater concern in determining a project's overall viability, particularly to developers that intend to retain ownership of the property after construction.

The amount of money lenders will provide in a permanent loan is tied to a percentage of the anticipated property value, as verified or accepted by the lender based on a third party appraisal. The amount of money a developer can borrow (loan amount) compared to the appraised value of the finished and stabilized property is known as the *loan to value* (LTV) ratio. As a simple example, if the developer was evaluating a project with an estimated final value of $10 million and the lender required a 75 percent LTV, the most money the developer could borrow is $7.5 million. The developer must provide the remaining $2.5 million needed or raise it from other sources, usually equity partners.

Lenders may be unwilling to provide the same level of funding when underwriting higher risk projects or projects sponsored by inexperienced or financially weak developers. In such cases (assuming the lender is still willing to make the loan), the LTV will be reduced to mitigate such risks. This reduces the amount of money committed by the lender and, thus, reduces the exposure should the loan not be repaid. It also reduces the size of the developer's mortgage payments, making it less likely that the developer will default, even if the project's final revenue stream is less than anticipated. However, reducing the LTV means the developer must source more equity commitments. If the same $10 million project was deemed risky by the lender, the LTV might only be 50 percent. In this case, the lender will provide $5 million and the developer will need to provide or raise an additional $5 million in equity.

While seeking opportunities in the earliest project stages, a developer will not know the exact value of a future, as-yet undetermined project. However, he/she can still estimate available financing terms based on past experience with lenders.

For example, if the developer's previous several projects were in the $10 to $15 million range, it is unlikely that the developer will suddenly obtain lending approval for a $100 million project. Similarly, the developer should evaluate what equity resources are available. If the most equity the developer has historically been able to raise is in the range of $2 to $5 million, he/she should not expect to source a significantly higher amount for the next project. This is not to say that developers do not eventually progress to larger, more sophisticated and expensive deals. Rather, building a track record of successful projects takes time and the size and scope of projects may increase incrementally over time. This cannot be rushed simply because the developer wants to pursue larger projects. Having a realistic idea of budget allows the developer to determine what opportunities are within a reasonable scope and prevents him/her from wasting time on deals that may ultimately be unobtainable. This will also help direct the developer's pursuits in the site search.

Needs versus Wants.   Development is an opportunity to shape the built environment. The lives of both future tenants and the surrounding community are influenced by how a site is laid out, how buildings are designed, and even by the view from the road as people drive past a completed development project. Understandably, at the start of a project, developers may have visions of creating a large landmark project, incorporating award-winning architectural design, including the latest environmentally friendly building innovations, and any number of other aspirations. While these are all laudable desires, they may, unfortunately, not be appropriate for a particular project's scope or budget.

In many cases, a developer's ambitions are less lofty, but equally infeasible, such as those tied to project budget or timeline. For example, imagine a developer who already owns a site and has identified a build-to-suit client whose business is expanding and needs a warehouse property ready for occupancy as soon as possible. To handle increasing capacity, the client is currently renting extra warehouse space at an inflated rate on a month-to-month basis. To capture this opportunity, the developer agreed to a firm 6-month delivery date with incentives for early delivery and now wants the development team to complete the project on an accelerated timeline. Even if the developer is willing to pass some of the early delivery incentives through to key members of the development team, accelerating the project simply may not be possible. Worse, efforts to do so could potentially lead to mistakes that actually lengthen the timeline. While the developer *wants* to have the warehouse completed in less than six months, what he/she actually *needs* is to be certain the project is ready for the client by the agreed date.

Separating things that are desirable from things that are truly and actually necessary can be challenging for developers, particularly if they are working with clients who arrive at certain preliminary decisions without the benefit of consulting with the development team. This can create future conflict with development team members who may be asked to facilitate the developer's ideas while also conforming to a fixed project budget and/or timeline. Engineers and other team members can help their developer-clients by providing limited early project support, even before being formally hired, to help the developer evaluate feasibility or consider other ways to achieve the developer's goals.

3.3.2. Accounting for Development Trends

As previously mentioned, developers are not likely to conduct full market studies while in the preliminary stages of seeking opportunities. However, it behooves the developer to have a general awareness of market trends that can influence the viability of a particular opportunity. Overall economic conditions, market demand, and the activity of other developers collectively influence the real estate market lifecycle. When several similar development projects are under construction in the same market area, the result may be a future oversupply of a particular product type. This translates to high vacancy and lower or stagnant rental rates, both of which can jeopardize a developer's projected returns.

For example, if the developer is aware that apartment vacancy rates have been slowly increasing and also that there are several new multifamily development projects are underway in a particular market, he/she may realize that vacancy rates are at risk for increasing dramatically once the new, additional apartment units are completed and become available for rent. Under such market conditions, the developer might prioritize looking for project opportunities in other markets or sites that are not tied to multifamily development. Conversely, if the developer is aware that large businesses have begun relocating to the market area, he/she might actively explore multifamily projects in anticipation of increased apartment demand for workforce housing.

Thus, before undertaking a project of any kind, the developer must evaluate the health of the real estate market and local economy not only in the present moment, but also through projections of future conditions. With the exception of presales/leases (which can be broken), future conditions—not current ones—will determine how quickly any new development can be sold or leased once it is delivered to the market.

3.3.3. Preliminary Project Planning

Although the design process typically begins after the developer has focused on a particular site (or sites), there are instances in which project planning discussions must begin before a site has been identified. For example, when a developer is working with a build-to-suit client, agreeing on a basic concept should be an early priority. Preliminary planning and design exercises facilitated by the assistance of a professional architect may help the client better refine its needs and explore options that had not been previously considered. Additionally, clarifying client needs may also help define implications for the site search so that the developer knows what size site will be required for the project. It is also important for the developer to know if the client is willing to colocate on the future site (or building) with other tenants. Doing so may help the developer pursue a larger range of sites and project types, but may not be desirable to a client that needs a single-user site for security or identity reasons.

CHAPTER CONCLUSION

This chapter focused on the initial stage of development planning. It outlined the types of preliminary considerations developers must be aware of when looking for project opportunities. It also described the different types of sites and properties that a developer may consider and outlined some of the challenges associated with each. For each type of site, examples of the different perspectives of members of the development team were presented, including the developer, the design team, attorneys, and lenders. Once a developer has made an initial determination about what type of project to pursue (Milestone 1: recognizing opportunity), he/she must then select and gain control of a suitable site for development, Milestones 2 and 3, respectively. The next chapter will discuss these milestones in the development process.

REFERENCES
1. 49 U.S. Code § 40103(b)—Sovereignty and use of airspace; 14 CFR 91.119—Minimum safe altitudes: General.
2. IRS Publication 946 (2016), How to Depreciate Property.
3. ICSC, Industry Insight series, "Physical Stores Dominant in Diversifying Grocery Landscape: Consumers Aware of New Options, Prefer Tangible Shopping Experience," September 8, 2017, available for download at https://www.icsc.org/uploads/t07-subpage/Physical-Stores-Dominant-Diversifying-Grocery-Landscape.pdf (accessed October 27, 2018).
4. North America Fit-Out Cost Guide Occupier Projects 2016/17 Edition, CBRE.

CHAPTER 4

SELECTING AND CONTROLLING THE SITE

Selecting a site (Milestone 2) and gaining control of the selected site (Milestone 3) are both critical milestones in the development process. Achieving these requires the active participation of several development team members. The developer's attorney and broker will be heavily involved, but members of the design team may also be called upon to help the developer make a preliminary evaluation of different sites. This chapter will present a discussion of the site selection process, the developer's preliminary financial analysis, and the different legal contracts involved in controlling a site.

4.1. THE SITE SELECTION PROCESS

Site selection is the process of identifying suitable sites for a development project. Entrepreneurial developers may be constantly looking for sites but, for many others, starting a site search requires a deliberate decision or triggering event. Regardless, once the basic parameters of a project have been decided (product type, approximate size/scope, etc.), a site search begins by selecting the specific geographic areas, or *markets*, in which the search will focus. For local developers, this may be an obvious choice, but large national developers may have the ability to select investment locations from across several markets in different states, counties,

or cities. Regardless of size, all developers should consider the impact of market conditions on their proposed projects and use this information to inform their decision making.

During the site search the developer will identify several potential sites and conduct a preliminary evaluation on each in order to select a preferred site (or sites) to pursue. Site searches can be quick or last several months, depending on the developer's preferences, available sites, and market factors. While some developers may have in-house specialists responsible for seeking out new opportunities, many rely either on their own market knowledge or hire a local commercial brokerage firm to assist with the search.

Commercial real estate brokers build their careers tracking sales, leases, new developments, and other market activity. Thus, they have specialized market knowledge as well as access to industry databases that track commercial properties, which are typically not included in residential multiple listing service (MLS) databases. Larger real estate service firms with brokerage divisions also usually have dedicated, in-house market research teams that can provide developers with a range of market data by geographical location and product type. In a typical site search, the developer's broker will identify available sites, contact the seller's broker to obtain site details, and help the developer negotiate terms for an acquisition. Note that most landowners also retain brokers to help market their properties for sale. Each broker represents the interests of their respective clients and helps facilitate the transaction.

Usually, the sites considered during a site search will be offered for sale; however, this is not always the case. Developers or their brokers may approach owners to solicit a sale or may partner with a landowner in order to create an opportunity.

4.1.1. Market Research

From the developer's perspective, *market research* refers to the process of gathering and evaluating information and data that is not site specific yet which is relevant to a proposed development project. This includes a review of both macro (general) and micro (local) market conditions. Macro market factors include national, regional, state, and/or county-level trends or conditions. Micro market factors are conditions influenced by demographics and other properties in the area near and around a site. Both macro and micro market factors can each impact a project. Collectively, market research can influence project-related decisions, inform the financial analysis of a proposed project, and assist with negotiations for land price.

Macro. All developers should be aware of macro considerations that have the potential to impact a project. Even if a local developer has witnessed several years of growth and prosperity in a particular neighborhood, it is not necessarily safe to assume that investing and developing in the area will inherently be successful; macro market conditions may simply not yet have revealed their influence. This is especially important to bear in mind given that development projects can take between one and three years (or more) to complete. Market conditions can change dramatically between the time a site is selected and when construction is finished. Thus, it is important for developers to consider both current and future market conditions. Three examples of macro market influences that can impact development are described next.

Example 1: Changing Demographic Trends

As both the Baby Boomer and Millennial generations continue to age, their needs and preferences evolve and continue to shape housing and multifamily rental markets. In many cases, the Boomers' demand for housing has started to shift from traditional single-family homes to multifamily rental units with full amenity packages. They will also likely need more healthcare facilities and various other types of senior living developments in the future. Demand for such facilities may not necessarily be distributed evenly across the country, or even within a region, as Boomers often move in retirement to warmer climates or to be near their adult children. Similarly, Millennials have demonstrated different housing preferences than previous generations, which suggest that homebuilders and apartment developers need to pay careful attention to trends associated with Millennial demand.

Example 2: Lifestyle Preferences

Retail customers are increasingly drawn to the convenience of shopping online and leave home in pursuit of entertainment rather than traditional shopping experiences. As a result, many traditional retail stores have been closing locations or downsizing at record rates, leaving a trend of growing vacancy in retail centers across the nation. In response, many retailers are changing their store concepts, which often changes how they use retail space, the size of the space they require, and can even influence how they identify and market their products. Before pursuing a traditional retail project, developers need to be aware of changes both in consumer preferences and retail tenant needs in order to create a successful project.

Example 3: Changes in Other Markets or Systems

The Federal Reserve System collectively serves as the central bank of the United States and is responsible for monetary policy, maintaining the stability of the U.S. financial system, and regulating financial institutions (among other things).[1] Decisions made by the Federal Reserve impact the U.S. economy generally and can also indirectly influence specific markets. In particular, increases in the Federal Fund Rate increase financial institutions' cost of funds, which leads to increases in residential mortgage rates. For homebuilders and multifamily developers, the actions of the Federal Reserve are extremely relevant. Changes in mortgages rates create financial incentives for people to either buy or rent, which affects demand for different housing products. Generally, if mortgage rates are high, more people will rent than purchase new homes; conversely, if mortgage rates are low, more people will leave the rental market to buy homes.

Micro. In addition to awareness of macro trends, a developer must also have an accurate understanding of market conditions surrounding a project site. This is important for determining a purchase price for the acquisition of the project site but also for assessing the conditions that will impact the project itself. These include rental rates, vacancy rates, overall demand for a particular product type, specific amenities demanded by the market, and other new development projects in the same area.

Arguably, the two most important micro market considerations for a developer to be aware of are submarket trends and comparable properties. *Submarkets* are smaller, distinct areas within a geographic market that are usually designated through convention by the local commercial real estate industry. *Comparable properties*, often referred to as "comps," are nearby properties that are substantially similar to the developer's own proposed project.

Submarkets within a county are often based on density and tend to be centered around municipal hubs or along transportation corridors; however, they can occur anywhere. Municipalities themselves often contribute to the creation of different submarkets through planning and zoning exercises that increase density in certain areas. Distinct

submarkets can also exist within large cities. For example, most cities have a *Central Business District* (CBD) submarket that is a corporate hub and often located at the city's center. CBDs tend to feature desirable Trophy and Class A office product and are often the focus of institutional investment or redevelopment activity because of continued corporate demand for the location.

Rental rates, standard amenities, and vacancy rates can vary tremendously between different submarkets, as can the cost to acquire sites. Each of these factors can influence the economics of a proposed project. Equally, different submarkets can have different demographic profiles, consumer habits, and growth patterns. The developer must be aware of all current and future submarket trends that can have direct impact on the success of a project.

Developers cannot simply focus on their own projects in comparison to submarket trends. They must also be aware of individual comparable properties that are substantially similar to their proposed project. It is important that the developer defines comparable properties accurately and objectively, or else he/she risks relying on inaccurate information and/or property comparisons. For example, two buildings of the same size with matching amenities may not be true comps if one benefits from a distinctly superior competitive geographic advantage, such as being located adjacent to a heavily used transit stop. The building with the better location will almost certainly be able to charge a higher rental rate than the building with the less desirable location.

When properly defined, comparable properties can serve as useful points of comparison for determining amenity packages or setting rental rates above or below market averages. However, they also represent competition as prospective tenants or homebuyers may be indifferent between a comparable property and the developer's own project. Further, it is important for the developer to be aware not only of existing comps but also of those that, much like the developer's own project, have not yet been built. Typically, the "pipeline" of future projects is tracked by the local Economic Development Office or Office of Planning and

Zoning. This information is usually publicly available and includes records of projects that are in the approval process as well as those that have obtained approval but may not yet be under construction. Reviewing the development pipeline can help a developer project the future competitive landscape around a particular site.

Market Considerations for Public Sector Projects.

Unlike private sector projects, public sector development projects are not under pressure to capture market demand or produce financial returns for investors. Therefore, market conditions are often not relevant in the same way for public sector projects. In most cases, a public need has already been determined and served as the catalyst to trigger the public sector project. However, if the government does not already own the land necessary to facilitate the project, the cost to acquire a site will be market dependent. Even if the public sector uses its powers of eminent domain to force a sale, as explained in Chap. 2, it must still pay a fair price for the property it acquires.

4.1.2. Site Considerations

Market factors are not the only thing developers must consider during the site search process. Obviously, site-specific considerations are extremely important. When evaluating potential sites, either individually or in comparison with other sites, a developer must consider location and access, size and shape, zoning, and cost. Each of these site-based factors can influence a development project, especially if the developer is working with a build-to-suit client.

Location and Access. Site location and access can be significant factors for development projects, although ideal requirements vary by product type. Further, location and access can either increase or decrease the desirability of the site for end users, which impacts the maximum rental rate (or sales price) the developer can charge. For example, conditions that are desirable for a retail project may be fatal for a residential community and vice versa.

Imagine that access for the Atherton Road site (Site A) was limited, as shown in Fig. 4.1. The site has essentially no frontage on Atherton Road and there is only one point of vehicular access with

no traffic light or dedicated turn lane. Atherton Road is a small, two-lane road, which suggests that traffic might not always be heavy; however, if it is a well-traveled route, particularly at certain times of day, it could still be difficult for visitors to access the site if they must turn from the far lane across traffic. Further, it is important to note that access matters in some states are heavily regulated by the state DOT. Therefore, even if a civil engineer can create a layout that appears to have suitable site access, the project may not be approved due to larger safety and congestion concerns. Both the developer and the design team must be aware that the realities of approval extend beyond simply finding a site design that works on paper.

Residential Use. Site A's limited access and location might be acceptable for a small residential community as occupants often prefer quieter environments away from main roads. The site effectively excludes "outsiders" and may help create a sense of privacy and community for residents. If a transit stop existed near the Atherton Road–Randall Road intersection, the site's vehicular access may be even less important, particularly if the developer and the design team can create appealing pedestrian access for residents at the northeast corner of the site where it touches Atherton Road. However, the site's location behind two retail centers also poses challenges. Dumpster odors and frequent early morning garbage collection, large delivery trucks, employees on smoke breaks, or other activity on the site's perimeter may be considered undesirable by residential users. To mitigate this, the developer and design team may be able to create sufficient privacy buffers between Site A and the rear of the existing retail centers through the use of fences, landscaping, and the orientation of the residential units themselves.

Office Use. As configured in this example, Site A is likely more suitable for a small office development. Unlike residential users, employees of an office tenant may appreciate the nearby retail centers if they provide convenient lunch venues. Further, office workers will not be bothered by frequent deliveries or garbage services because they are not likely to be at work before or after customary business hours when such activities typically occur. The site's

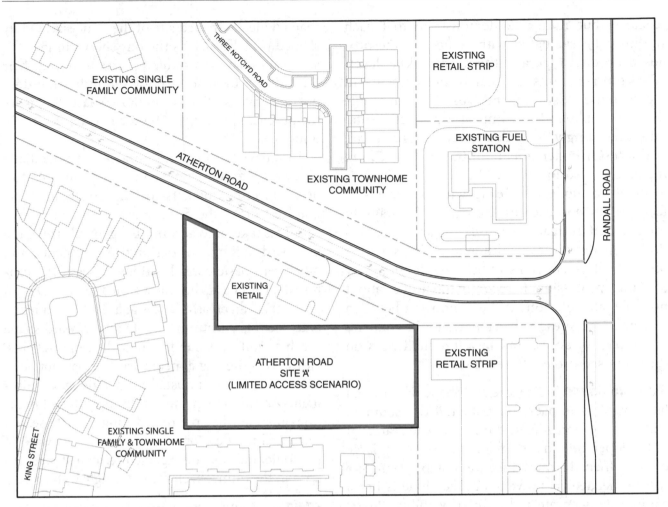

FIGURE 4.1 Atherton Road site with limited access.

limited visibility on Atherton Road may or may not matter to an office tenant. Assuming that both adjacent retail centers are comprised of single story buildings and that a two or three story office building is permitted on Site A, the developer may be able to offer building signage that would be visible from the road. However, office tenants, particularly smaller firms, do not always warrant building signage and would need to accept lack of visibility on the site. With respect to site access, office employees driving to the site every day will be familiar with its location, but the limited access and visibility may make it difficult for the firm's clients to find the building. This may make the site less desirable to certain office users that rely on client visits.

Retail Use. Despite its potential for a residential or office project, Site A in this configuration would be disastrous by the standards of most retailers because of the complete lack of roadside visibility and limited access. Most retail tenants are extremely sensitive to these factors and prefer high-visibility sites with easy access from a heavily traveled street. Indeed, many retail tenants will include the daily average trip count on adjacent roads as part of their decision making when evaluating new store locations. While the developer might be able to negotiate an access easement with the existing retail centers such that cars could access Site A through the other retail parking lots, there is very little that can be done to address the site's limited visibility.

There are an infinite number of potential location and access conditions a developer can encounter during a site search. The purpose of this section is not to attempt and describe each scenario, but rather to demonstrate that each site must be evaluated with respect to the proposed product type and the needs of the developer's future tenants. Ultimately, site location and access issues exist on a spectrum from desirable to undesirable and can impact a project's rental rates accordingly. Note that lower rental rates are not necessarily detrimental if the cost to acquire the land is also lower and the developer intends to create a modest building without expensive features. In some cases there may be ways to improve or mitigate location and access conditions. The design team, although perhaps not yet officially engaged, can still help the developer by offering professional recommendations about possible site solutions. When choosing between two or more sites, a developer should compare the impact of access and location challenges on the proposed project for each site.

Size and Shape.   The size and shape of a site can have obvious impacts on potential development and must be considered during the site search. If a developer intends to create an extremely large project, then, in most cases, a suitably large site will be needed. However, site size alone is not a completely accurate indicator of viability. Physical features such as shallow rock, asbestos soils, steep slopes, and bodies of water can significantly reduce the developable area of a site. As a simplified example, if a stream flows through a 5-acre property, resulting in two of the acres being defined as stream, wetlands, and resource protection areas, then only three acres of the site may actually be available for development. Such site conditions can quickly change the perceived value of a site while also impacting the project and potentially increasing construction costs.

A site must also be shaped in such a way to allow a useable configuration for the development project. For a development of any size, a developer will always prefer to avoid paying for excess or unusable land, especially in urban markets where land prices can be extremely expensive. Although many oddly shaped sites can still be developed, the resulting layout will be inefficient and site utilization will be poor. This means that the site will yield far fewer residential units or square feet of space

than an ideally shaped site of the same size. Oddly shaped parcels, such as the one shown in Fig. 4.2, sometimes have discounted prices to reflect their limited use. Even so, if poor site utilization does not generate sufficient rentable space to justify the purchase price, then the site may be effectively unusable.

To quickly evaluate a desirable but potentially awkward site, developers may rely on architect or engineering contacts to help create an informal "test fit" of the project layout on a particular site. This takes into account the general size of the proposed building(s), parking requirements, and other very basic considerations to see if the proposed project can be placed on the site. Design teams that can creatively layout a site to accommodate necessary features despite an awkward shape can help unlock value for their developer-clients. However, a site that appears useable may not be in actual fact due to unusual jurisdictional requirements or unforeseen site conditions. These limitations may not be apparent during the site search stage and the developer must always be aware of the potential for unexpected circumstances to arise in the future, even for seemingly viable sites.

Zoning.   A site's zoning designation can have tremendous impact on a proposed project. When looking at site size and parcel boundaries, such as shown on a tax map, a site may look sufficiently large enough to accommodate the developer's vision for the project. However, size and boundaries are not accurate indicators of whether or not the site will be able to accommodate the development program. Zoning ordinance requirements will often reduce the area of the site that can be developed, sometimes significantly. Regulatory conditions such as building setbacks, open space requirements, environmental protection areas, historic overlay areas, floodplains, easements, necessary road frontage and access requirements, and other site conditions can all limit how much of a site can actually be developed. Additionally, whenever possible, early efforts should be made to verify that the proposed project and site area can accommodate the necessary infrastructure. Regulatory requirements will also often govern the size, type, and location of utility and stormwater management

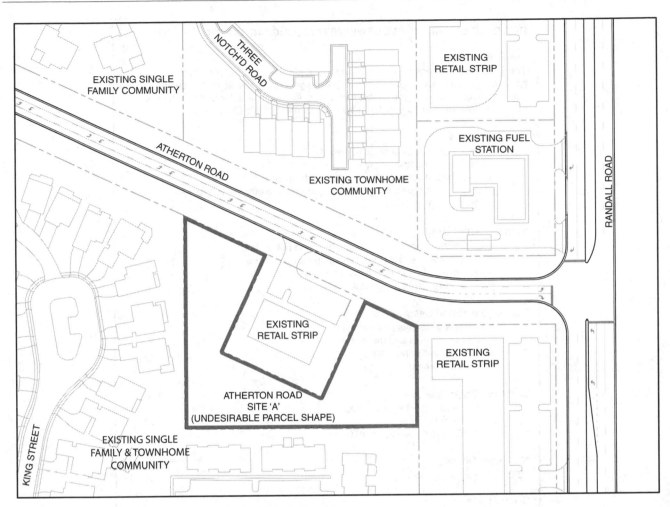

FIGURE 4.2   Atherton Road site with inefficient shape.

systems, which often occupy large areas of a site. If improvements (buildings, infrastructure, or otherwise) are proposed too close to an adjacent site, it can trigger the need to obtain permission or easements from adjacent site owners, which introduces a new risk for the developer as he/she must now negotiate with another party in order for the project to be successful. If consulted early enough, the developer's site engineer can advise on design and constructability concerns and identify potential risk in the early phases of a project. The task of evaluating buildable area should be performed as early as possible to identify the site constraints and coordination with the site engineer and public officials can uncover many restrictive conditions before a significant investment has been made on a particular site. However, during the site selection process, it is not always feasible for the developer

to begin incurring design-related costs to investigate these issues in detail, which can be frustrating for members of the development team who are hired later in the process only to discover limiting site conditions on the developer's chosen site.

To demonstrate some basic examples of the impact of zoning on development, we will explore three common scenarios using the original 2-acre Atherton Road (Site A) infill site with an R-12 medium density residential zoning designation, as shown in Figs. 4.3 and 4.4. Using this information, we can track the options facing a developer interested in building on the site. Note, this section is primarily concerned with providing simplified examples of zoning implications but does not address the zoning approvals process, which will be described in Chap. 6. Further, the following scenarios make certain market, site design, and

---

R-12 RESIDENTIAL DISTRICT, TWELVE DWELLING UNITS/ACRE

Purpose

The R-12 District is established to provide for a planned mixture of residential dwelling types at a density not to exceed twelve (12) dwelling units per acre; to allow other selected uses which are compatible with the residential character of the district; and otherwise to implement the stated purpose and intent of this Ordinance.

Permitted Uses

1. Dwellings, single family attached
2. Dwellings, multiple family
3. Dwellings, mixture of those types set forth above
4. Churches, chapels, temples, synagogues, and other such places of worship
5. Public uses

Special Permit Uses

1. Convents, monasteries, seminaries, and nunneries
2. Subdivision and apartment sales and rental offices
3. Temporary dwellings or mobile homes
4. Temporary farmers' markets

Special Exception Uses

1. Child care centers and nurseries
2. Cultural centers and museums
3. Independent living facilities
4. Funeral homes

Lot Size Requirements

1. Minimum district size: 2 acres
2. Minimum lot area:
   A. Nonresidential uses: 10,000 sf

Maximum Density

12 dwelling units per acre

Open Space

25% of the gross area shall be open space

Parking Requirements

2 dedicated spaces per dwelling unit with 1 additional visitor space per dwelling unit

FIGURE 4.3   Sample residential district regulations. (*Source: Based partially on Fairfax County, VA codes.*)

financial assumptions in order to highlight zoning issues, but will not delve into an explanation of these topics themselves.

Scenario 1: The Developer Wants to Build a Small Townhome Community

The necessary zoning is already in place and the developer can build up to 24-townhome units (single-family attached dwellings) by right. However, Atherton Road cuts into the top portion of the site, making the shape inefficient. Due to the odd shape, there is not enough room to build all 24 units and

meet both the open space requirement and the prescribed off-street parking requirement. The developer's civil engineers are able to accommodate required parking for all 24 units but can only achieve a 23 percent open space factor. Even though this is close to the required 25 percent open space requirement, it is not compliant. The developer has several choices: (1) underutilize site density and build one less unit for a total of 23 townhomes, (2) submit a request to deviate from the zoning requirements, (3) explore building a multifamily apartment building, which is also permitted by right, instead of building

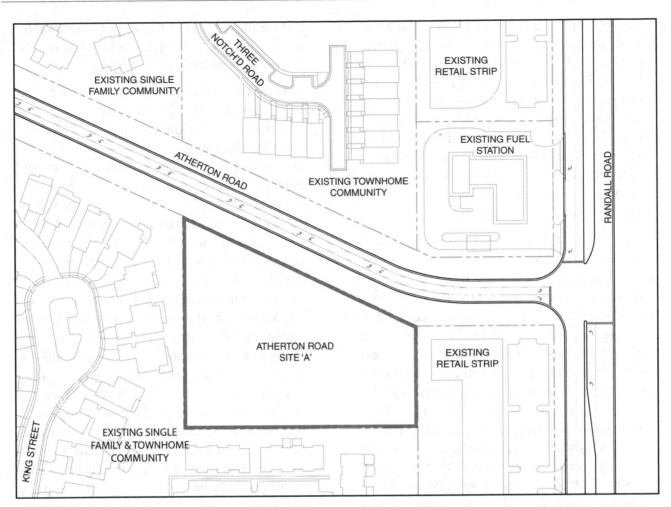

FIGURE 4.4 Atherton Road site.

townhomes, or (4) decide not to purchase the site. For purposes of this scenario, we assume that the developer does want to buy the land but is a townhome builder with no interest or competency in building apartments.

If the developer builds 23 units instead of 24, he/she could reallocate the unused square footage of the 24th unit and use it to meet the open space and parking requirements of a 23-unit community. In this sense, forgoing the 24th unit allows the developer to move forward with the project. However, the developer does not capture any appreciable cost-savings by building only 23 units instead of 24. In fact, the majority of the project cost will still be incurred whether the developer builds the 24th unit or not. Thus, it behooves the developer to maximize density in order to

spread costs across the sale (or rental) of as many units as possible. Therefore, given the choice, the developer in this scenario would prefer to seek permission to build all 24 units.

As explained in Chap. 2, it is possible to deviate from zoning ordinances in certain situations through a request for either a variance or a special exception. Although use-variances exist in certain jurisdictions, the most common form of variances are tied to physical limitations of a site that make it difficult or impossible to meet existing zoning requirements, such as setback distances. Note that a developer's chances of success for being granted the variance depend on the specifics of the variance being requested. In this scenario, the developer is asking for a minor reduction in the open space requirement.

This request is unlikely to impact the quality of the community, the surrounding area, or the overall integrity of the comprehensive plan, so we will assume that the variance is granted. However, if Site A were smaller and even more awkwardly shaped, the developer might have to ask for a variance to reduce the open space requirement significantly. If the resulting development would not be consistent with a suburban environment, such a request would likely be denied.

Note that, even if the developer is successful in pursuing a variance, doing so costs the developer time and money. The cost may be significant to prepare site plans demonstrating that 23 percent is the maximum achievable open space and it could take between three and nine months for the developer's request to be considered by the Office of Planning and Zoning. Although we stipulated that the developer in this scenario *did* want to purchase the land, not all developers would make the same business decision. If a developer did not want to engage with the challenges posed by Site A in this scenario, the site would be deemed nonviable and the developer would continue the site search.

Scenario 2: The Developer Wants to Build a Childcare Center

The developer in this scenario is not a homebuilder, but rather an entrepreneur working with a daycare operator client for a build-to-suit project. The client likes Site A because of all the homes in the surrounding area and believes it will be a successful business location. If the developer is able to build a daycare center on the site, the daycare client will sign a long-term lease that will provide the developer with a rental income stream for many years. However, per the zoning regulations, daycare is not a by right use. In order to receive permission to build the center, the developer will need a special exception. As previously discussed, special exceptions can be granted in cases where the proposed use may be generally consistent with the intent of

the zoning district but have potential impact on surrounding properties or the community that require special consideration.

In this case, a daycare use will increase traffic more than the by right use of 24 townhomes. The required residential parking requirement for the townhomes is spaces for 72 cars. However, a large daycare center can accommodate as many as 200 children. In a best-case scenario, the 200 children are comprised of sets of siblings, translating to 100 cars making drop-offs and pickups each day; in a worse-case scenario, each of the 200 kids are single children without siblings, requiring 200 separate cars making drop-offs and pickups each day. Either case represents an increase in trips beyond what is anticipated under the site's permitted residential use. Further, unlike a residential community where people may have staggered schedules, daycare drop-off and pickup times are fairly rigid. This means that as many as 200 cars could try to access the site within a short time frame twice a day: morning drop-off and evening pickup. This could certainly have a negative impact on the surrounding area, for example, by increasing traffic congestion if cars exit the site and then begin to back up as they wait to make a turn through the Atherton Road–Randall Road intersection. Note that these estimates do not include the cars driven by the daycare center employees.

Traffic may not be the only concern associated with this special exception request. A large daycare can also cause more noise than might otherwise be anticipated in a residential zone. While children playing is, of course, expected and appropriate in residential areas, there are not normally large groups of children in the same area every day, even in neighborhood parks. If multiple classes of children from the daycare all use the playground at the same time, it could be disruptive to certain adjacent uses. While this may seem unlikely, it is, nonetheless, the job of the zoning commission to make sure the issue is given due consideration. Note these are only two simple examples of concerns that might be associated with a day-care special expectation in a residential zone. Different zoning categories will have a range of different specially excepted uses and the unique circumstances of each special exception request will have to be taken into account as part of each evaluation.

In order to receive approval for the special exception in this daycare scenario, the developer could try to assuage traffic concerns by agreeing to move the site access further from the intersection. He/she might also try to negotiate an agreement to have access through the parking lot of the adjacent retail site, which would funnel daycare traffic through an already existing route. Equally, the zoning board might make the special exception contingent on such solutions, even if they are not offered by the developer. Much like the variance in Scenario 1, the developer's decision to pursue a special exception will require a certain investment of time and money.

Scenario 3: The Developer Wants to Build a 40-Unit Apartment Building

In some cases, a developer may want to build something that does not comply with the existing zoning for a site. This could be a completely different use than what is permitted, but often the developer simply wants to increase the density of the permitted use beyond what is allowed by right. In Scenario 3, assume the owner of Site A increased the sale price such that the developer must now build 40 residential units in order to cover costs and make enough profit to provide sufficient returns to equity investors. However, the only way for the developer to obtain permission to build a 40-unit apartment building is to officially change the site's existing zoning. In this case, the zoning would need to be changed from the existing R-12 to a higher density R-20 zoning, which would allow for 20 residential units per acre.

After determining the necessary zoning, the developer must consult the comprehensive plan to evaluate the compatibility or conflict between the plan and the proposed new site zoning. He/she must then assemble a comprehensive package of information to be provided as part of the rezoning application addressing how the proposed change might impact the surrounding area. The rezoning application is subject to intense scrutiny. It will be the subject of technical reviews, planning staff evaluation, review by other government agencies (such as the transportation administration), as well as public hearings. Of these, the public hearings are arguably the most challenging as affected residents respond emotionally to both real and perceived impacts on their neighborhoods. Ultimately, the burden falls to the developer to make the case that the proposed rezoning is appropriate and beneficial for the community.

The rezoning process is an extensive one that does not fit many developers' level of risk tolerance. It is entirely possible that an entire year (or more) will pass between the time the developer initially files a rezoning application and when the final decision is issued by the relevant legislative body. Further, the plan that is ultimately approved may be different from the developer's original proposal because changes are often required in order to satisfy different stakeholders.

For purposes of our Scenario 3, the developer will need to make a business decision about the cost and risk of trying to rezone Site A. Note that the seller is unlikely to wait through the lengthy rezoning processes and will sell the site to someone else if possible. The developer has two ways to retain control of the site: (1) move forward with the purchase and acquire the site without knowing whether or not it can be rezoned, or (2) use an option contract (or other contractual vehicle) and pay the owner to "reserve" the site for a certain amount of time while the developer pursues the rezoning. The first option is too risky for the developer and, given the already increased price of acquisition, the second option may be too expensive. In this scenario, the zoning challenge causes the developer to decide not to pursue the site.

Note that some zoning regimes accommodate a special mechanism that allows developers to transfer density between unrelated sites. This is achieved through a specially legislated commodity known as a *transferable development right* (TDR) as described in Chap. 2. Through the use of TDRs, the owner of one property can sell unused density to the owner of another property, who can then develop a larger project than he/she might have been able to by right. TDRs serve as a planning tool that can help preserve lower density development in protected parts of a community, such as historic areas, without penalizing owners economically for development restrictions imposed in those areas. Simultaneously, TDRs can incentivize higher density development in targeted growth areas. These are often known, respectively, as "sending" and "receiving" areas within the comprehensive plan. Had Site A been TDR-eligible, the developer might have been able to move forward with the 40-unit apartment building without applying for a rezoning.

Cost. The cost to acquire a site, also known as the *land cost* or *purchase price*, has a direct impact on total project costs. While the purchase price is often negotiable to some degree, the value of a particular parcel of land is determined by a variety of factors. Location is a critical factor of land cost. Sites located in a highly desirable area will almost always be more expensive than equivalent sites located in less desirable areas. A higher land cost in such instances may not be fatal to a project if, as a result of the desirable location, the developer can increase rental rates for the project. Scarcity is another factor that impacts land values. When only a limited number of sites are available in a certain area or with certain characteristics, the cost to acquire one will be greater than in markets with abundantly available land. Urban environments often combine location and scarcity constraints, resulting in particularly high land costs compared to suburban or rural markets.

Existing entitlements can also increase land value. When entitlements are already in place, a developer can begin building without needing to seek certain zoning or land use approvals. This accelerates project time lines while also reducing the risk that approvals will be denied or delayed. Indeed, some specialty developers focus on taking sites through necessary zoning changes and the entitlement processes in order to create value. These developers may not actually build on the sites themselves, but rather sell the land to another developer for a premium. Focusing only on entitled sites can be a priority of a site search, depending on a developer's preferences and rezoning risk tolerance.

There are also several other factors that can reduce site value. These include awkward site shape, poor access, the existence of easements or restrictive encumbrances, and contamination. While sites suffering from any of these conditions may be more affordable, the cost to develop them may be greater due to the need to mitigate underlying problems (if possible). Note that some undesirable site conditions, such as contamination, may not be readily apparent during the site search.

Site Considerations for Public Sector Projects. With the exception of zoning, many site considerations are similar between public and private development projects. Site size and shape must be conducive to the proposed public sector project, regardless of whether that project is a new utility or police station. Cost can equally impact feasibility, particularly for budget-constrained public projects or with respect to the cost to acquire off-site easements. Location and access concerns also exist for the public sector, although they may be based on different criteria than private sector developments.

4.2. "BACK OF THE ENVELOPE" FINANCIAL ANALYSIS

In order for a potential development opportunity to be viable, it must meet several financial hurdles. At the early stages of evaluation, however, it is not worth the time or cost for a developer to perform a complete financial analysis on every potential site. Indeed, it may not even be possible given the many unknowns that still exist during the site search stage. Rather, the developer will do a simplified analysis on promising sites to determine if an opportunity might be feasible. This is classically referred to as a "back of the envelope" calculation because it is so basic that it can be written out informally on whatever scratch paper is readily available, such as the back of an envelope. By examining whether estimated value exceeds cost, this method provides a quick way to evaluate whether or not a site warrants further consideration.

For purposes of demonstrating a "back of the envelope" analysis in the following sections, we return to the scenario in which the Atherton Road site (Site A) is zoned for low-density office use. In Chap. 3 we used this scenario as part of a hypothetical exercise to demonstrate project scope and budget. However, we now assume that Site A has actually been identified as part of the site search and is being considered for the proposed project. The developer no longer has to make general assumptions about site size, allowable floor area ratio (FAR), land cost, or rental rates. Instead, he/she can account for specific site considerations as well as market information in the "back of the envelope" analysis. The following information will be used in this example throughout the following sections unless otherwise stated:

4.2.1. Refining Project Cost Estimates

Recall that when using this office scenario in Chap. 3 we calculated project cost using hard costs of $100 per square foot (psf), soft costs at 10 percent of hard costs, and a 5 percent contingency factor. We will continue to use these same assumptions but now, with the actual site size and purchase price known, we can update the calculation of total project cost as shown in Fig. 4.5.

With the total estimated project cost calculated, now including the cost of the land, the developer must also calculate an estimate of project value in order to determine whether or not the project value outweighs costs. Obviously, there must be more value created than cost incurred in order for the project to be worthwhile for the developer.

4.2.2. Calculating Net Operating Income and Estimating Project Value

In order to arrive at a value for the project, the developer needs to estimate how much revenue the project would generate during a typical year if the building was completed, leased to tenants, and operating. This measure of cash flow for a stabilized asset is known as the building's *net operating income* (NOI). NOI is important for several reasons. It can be used to estimate project value

Site A

Site Considerations

   Zoning: allows low-density office development by right

   FAR: 0.4

   Site size: 1.25 acres

   Location/access: highly desirable

   Size/shape: allows for full FAR development

   Price: $500,000 per acre (or $625,000 total)

Market Considerations

   Office rental rate: $17.50 psf average in the submarket for new buildings

   Office vacancy rate: 5%

   Office cap rate: 6%

Site A		
Calculating Hard Costs		
	21,780	Size of building in square feet
×	$100	Hard cost psf
	$2,178,000	Total estimated hard costs
Calculating Soft Costs		
	$2,178,000	Total estimated hard costs
×	10%	
	$217,800	Total estimated soft costs
Calculating Construction Costs		
	$2,178,000	Total estimated hard costs
+	$217,800	Total estimated soft costs
	$2,395,800	Total estimated construction costs
Calculating Total Estimated Project Cost		
	$2,395,800	Total estimated construction costs
+	$119,790	5% contingency (of construction costs)
	$625,000	Land cost
	$3,140,590	Total estimated project cost

FIGURE 4.5 Site A: project cost calculations.

as well as help the developer determine the maximum purchase price he/she is willing to pay for a particular site. NOI also represents the amount of money available to cover mortgage payments, compensate equity investors, and provide a revenue stream for the developer (or investor if the project is sold).

In order to continue with the "back of the envelope" analysis and arrive at NOI, the developer must make several simple calculations. First, the maximum rental income the property is capable of achieving must be determined. This measure, referred to as *potential gross income* (PGI), reflects the total potential revenue the developer could receive if 100 percent of building space were to be leased at the advertised, asking rental rate. To calculate PGI, the developer relies on market information, which in this case suggests that average office rental rates are $17.50 psf. With this information, PGI is calculated as shown in Fig. 4.6.

Site A	
Calculating Potential Gross Income (PGI)	
	21,780 Size of building in square feet
×	$17.50 Rental rate psf
	$381,150 Potential Gross Income (PGI)

FIGURE 4.6   Site A: calculating PGI.

The entire building is unlikely to be completely leased at every point during its operation, however, and the developer must account for a vacancy factor. Again relying on market data, in this scenario we know that the average vacancy rate for office buildings in the submarket is 5 percent (of PGI). Accounting for the inevitable vacancy of space or loss of rental income due to collection problems results in the property's *effective gross income* (EGI). Note that if the developer were expecting to collect additional rent from other sources, such as charging for parking or renting roof space for telecommunications antennas, those additional income streams would be added in the calculation of EGI. For purposes of this example, we assume that the developer will have no other revenue streams and an EGI as calculated in Fig. 4.7.

Site A	
Calculating Effective Gross Income (EGI)	
	$381,150 Potential Gross Income (PGI)
−	$19,058 Vacancy / Collection loss
	$362,092 Effective Gross Income (EGI)

FIGURE 4.7   Site A: calculating EGI.

Note that calculations are shown rounded to the nearest whole number as an accurate representation of the "back of the envelope" exercise, which often includes minor mathematical discrepancies as a result. In some cases, developers will even round to the nearest hundred-thousand dollar amount for simplicity. In these early financial estimates, determining the general order of magnitude of revenue/costs is the priority rather than achieving 100% accuracy in calculations. While many input factors are now known, a multitude remain unknown and accurate financial projections are not possible at this stage, with or without rounding inconsistencies. This casual approach may be very uncomfortable to different members of the design team, but it is an important part of understanding the process of their developer-clients.

Having determined the estimated amount of income he/she can expect to receive from the project, the developer must now account for the cost to operate the property. *Operating expenses*, often abbreviated as "OpEx," are those costs incurred in order to operate the property. Collectively, OpEx includes costs associated with maintenance, utility bills (if paid by the owner), property management fees, taxes, and services like landscaping or snow removal. During the final pro forma analysis, the developer may get actual service bids and break out each of the different operating expense components separately. However, for the "back of the envelope" calculations, the developer is only concerned with an estimate of cost and will most likely rely on personal knowledge or use industry averages to calculate OpEx. It is worth noting that operating expenses vary by product type. For many property types, they are expressed as a cost per square foot but can also be estimated as a percentage of either PGI or EGI. For purposes of this example, we will assume that operating expenses are equal to $6 psf and calculated as shown in Fig. 4.8.

Site A	
Calculating Operating Expenses (OpEx)	
21,780	Size of building in square feet
× $6	Approx. cost psf to operate the building
$130,680	Operating Expenses (OpEx)

FIGURE 4.8   Site A: calculating operating expenses.

Once the values for PGI, EGI, and OpEx are determined, the developer can finally calculate the project's annual NOI as shown in Fig. 4.9.

Note that NOI is a quick "snap shot" estimation of the potential gain or loss from building operations in a 1-year period. In both the "back of the envelope" analysis and also in a full pro forma analysis, NOI is always an estimate because actual conditions cannot be known until a particular year of operations is over. Nonetheless, NOI is a useful tool for making comparisons between different projects and also as a tool to estimate project value.

Site A	
Calculating Net Operating Income (NOI)	
$381,150	Potential Gross Income (PGI)
– $19,058	Vacancy / Collection loss
$362,092	Effective Gross Income (EGI)
– $130,680	Operating Expenses (OpEx)
$231,412	Net Operating Income (NOI)

FIGURE 4.9   Site A: calculating NOI.

To calculate an estimate of project value, the developer applies a capitalization rate to the NOI. *Capitalization rates*, or "cap rates," are simply ratios of the purchase value of completed and operating projects compared to their NOIs. While they are helpful indicators of market trends for property value, cap rates can be inaccurate and contrived and should not be relied upon by the developer for final decision making. However, they serve the developer well in the early stages of seeking

opportunity by providing a quick indication as to whether or not a particular property is over- or undervalued.

As per the market research in this case, the average cap rate across several transactions in the market surrounding the Atherton Road site is 6 percent. The developer divides the project NOI by 6 percent, as shown in the formal below, to arrive at a rough opinion of value as shown in Fig. 4.10.

$$\text{Estimated project value} = \frac{\text{NOI}}{\text{Cap rate}}$$

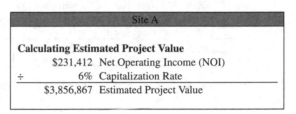

Site A	
Calculating Estimated Project Value	
$231,412	Net Operating Income (NOI)
÷ 6%	Capitalization Rate
$3,856,867	Estimated Project Value

FIGURE 4.10   Site A: calculating estimated project value.

According to a back of the envelope cap rate analysis, the total project value is $3,856,867. In other words, were the project to be built and operating, the stabilized building and the cash flow it produces would be valued at $3,856,867— or roughly $3.85 million—based on these early assumptions.

The developer can now compare the project value to total cost. Essentially, sites may be worth the developer's time to explore further if the estimated total project cost (including purchase price of the land) is less than or equal to the estimated total project value. As previously stated, more value must be created than cost incurred in order for the project to be worthwhile for the developer. In this case, Site A passes the test:

$3,140,590 < $3,856,867

(project cost < project value)

Complete side-by-side "back of the envelope" calculations for project cost and value are shown in Fig. 4.11.

Site A - Project Cost		
Calculating Hard Costs		
	21,780	Size of building in square feet
×	$100	Hard cost psf
	$2,178,000	Total estimated hard costs
Calculating Soft Costs		
	$2,178,000	Total estimated hard costs
×	10%	
	$217,800	Total estimated soft costs
Calculating Construction Costs		
	$2,178,000	Total estimated hard costs
+	$217,800	Total estimated soft costs
	$2,395,800	Total estimated construction costs
Calculating Project Cost		
	$2,395,800	Total estimated construction costs
+	$119,790	5% contingency (of construction costs)
	$625,000	Land cost
	$3,140,590	Total estimated project cost

Site A - Project Value		
Calculating Potential Gross Income (PGI)		
	21,780	Size of building in square feet
×	$17.50	Rental rate psf
	$381,150	Potential Gross Income (PGI)
Calculating Operating Expenses (OpEx)		
	21,780	Size of building in square feet
×	$6	Approx. cost psf to operate the building
	$130,680	Operating Expense (OpEx)
Calculating Net Operating Income (NOI)		
	$381,150	Potential Gross Income (PGI)
−	$19,058	Vacancy / Collection loss
	$362,092	Effective Gross Income (EGI)
−	$130,680	Operating Expenses (OpEx)
	$231,412	Net Operating Income (NOI)
Calculating Project Value		
	$231,412	NOI
÷	6%	Capitalization Rate
	$3,856,867	Total estimated project value

FIGURE 4.11 Site A: project cost and project value calculations.

These simple calculations can be completed for any or all of the sites identified in the developer's site search to help evaluate feasibility. Assume now that, in addition to Site A, the developer's site search yielded a second potential site in the same submarket. This second site is located on Baldwin Court (Site B) and is described by the following information and calculations in Fig. 4.12.

Site B

Site Considerations

Zoning: allows low-density office development by right

FAR: 0.4

Site size: 1 acre

Site B - Project Cost		
Calculating Hard Costs		
	17,424	Size of building in square feet
×	$100	Hard cost psf
	$1,742,400	Total estimated hard costs
Calculating Soft Costs		
	$1,742,400	Total estimated hard costs
×	10%	
	$174,240	Total estimated soft costs
Calculating Construction Costs		
	$1,742,400	Total estimated hard costs
+	$174,240	Total estimated soft costs
	$1,916,640	Total estimated construction costs
Calculating Project Cost		
	$1,916,640	Total estimated construction costs
+	$95,832	5% contingency (of construction costs)
	$600,000	Land cost
	$2,612,472	Total estimated project cost

Site B - Project Value		
Calculating Potential Gross Income (PGI)		
	17,424	Size of building in square feet
×	$17.50	Rental rate psf
	$304,920	Potential Gross Income (PGI)
Calculating Operating Expenses (OpEx)		
	17,424	Size of project in square feet
×	$6	Approx. cost psf to operate the building
	$104,544	Operating Expense (OpEx)
Calculating Net Operating Income (NOI)		
	$304,920	Potential Gross Income (PGI)
−	$15,246	Vacancy / Collection loss
	$289,674	Effective Gross Income (EGI)
−	$104,544	Operating Expenses (OpEx)
	$185,130	Net Operating Income (NOI)
Calculating Project Value		
	$185,130	NOI
÷	6%	Capitalization Rate
	$3,085,500	Total estimated project value

FIGURE 4.12 Site B: project cost and project value calculations.

Location/access: acceptable

Size/shape: allows for full FAR development

Price: $600,000 per acre (or $600,000 total)

Market Considerations

Office rental rate: $17.50 psf average in the submarket for new buildings

Office vacancy rate: 5%

Office cap rate: 6%

Although the total cost to undertake the proposed project on Site B is less than the cost of the project on Site A, so too is the total estimated project value. However, project value still outweighs project cost for Site B. For our scenario, we assume Site A is the developer's preferred site because of the greater value of the proposed project and the site's highly desirable location and access. Site B is still an acceptable site, but the developer considers it a "backup" option.

In instances where costs outweigh value, there are several options available to the developer: (1) deem the site/project nonviable and continue looking for other sites, (2) attempt to reduce costs by negotiating a lower purchase price for the land, (3) consult with the design team to explore creative ways to overcome any limiting site layout factors, (4) consider the cost and viability of changing zoning to increase density, or (5) reconsider the proposed project to see if/where costs can be reduced. If value and cost estimates are acceptable for a desirable site, there are additional "back of the envelope" calculations the developer must complete in order to estimate the cost of debt financing and project returns.

4.2.3. Evaluating Basic Debt Financing Requirements

The developer now has a more accurate, site-specific understanding of total project cost and value for both Site A and Site B. He/she also has an estimate of annual project cash flow from NOI. However, it is important to remember that NOI does not account for loan payments the developer will have to make in exchange for lender financing. At this stage in the "back of the envelope" analysis, the developer must evaluate basic loan requirements to determine if the proposed project on either site is viable from a financing perspective.

By the time the site search has started, the developer should already have done an initial evaluation of what project scope and financing options are realistic, as outlined in Chap. 3. Those preliminary considerations can now be applied to assess the debt and equity implications of pursuing a project on either Site A or B. Note that the developer will need to obtain a construction loan in order to fund the actual construction of a project on either site; however, construction loans are temporary, often interest-only, loans that are replaced by a permanent loan once construction is finished and the property is rented and stabilized. Because permanent loans replace construction loans and will remain in place during the building's operating lifetime (or until a refinance), it is the viability of the proposed projects with respect to permanent financing that the developer wishes to evaluate at this stage.

The loan amount available to the developer under a permanent loan is determined by applying a *loan to value* (LTV) factor. LTV is the ratio of the amount of money a developer can borrow compared to the value of the finished and stabilized property, or

$$\frac{\text{Loan amount}}{\text{Total estimated value}} = \text{LTV}$$

or

$$\text{Total estimated value} \times \text{LTV} = \text{loan amount}$$

During the "back of the envelope" calculations, the developer will use the project value based on NOI, as calculated in the previous section. Note that loan-to-cost ratios are often evaluated by both developers and lenders for some projects, but will not be the focus of this chapter. To continue with our example, we assume the developer is confident he/she can obtain a loan with the following conditions for either site:

Financing Assumptions

LTV: 70%

Interest rate: 5.5%

Term: 30 years

A 70 percent LTV applied to the proposed projects for both Site A and Site B yields the loan amounts shown in Fig. 4.13.

Site A	
Calculating Loan Amount	
	$3,856,867 Estimated Project Value
×	0.7 LTV
	$2,699,807 Loan Amount

Site B	
Calculating Loan Amount	
	$3,085,500 Estimated Project Value
×	0.7 LTV
	$2,159,850 Loan Amount

FIGURE 4.13 Loan calculations for Site A and Site B.

Site A	
Calculating Equity Contribution	
	$3,856,867 Estimated Project Value
–	$2,699,807 Loan Amount
	$1,157,060 Equity Required

Site B	
Calculating Equity Contribution	
	$3,085,500 Estimated Project Value
–	$2,159,850 Loan Amount
	$925,650 Equity Required

FIGURE 4.14 Equity calculations for Site A and Site B.

The developer must make an equity, or cash, contribution to the project equal to the difference between the loan amount and the estimated project value. The amount of equity required for each proposed project is calculated in Fig. 4.14.

In this case, the developer will need to contribute over $1 million to pursue Site A and less than $1 million to pursue Site B. If the developer does not have this amount of cash available, he/she will need to raise the required amounts by seeking equity investors to contribute funds toward the project. For our example, we assume the developer can raise a maximum of $1 million for the proposed project. This makes the difference in the equity requirements between Site A and Site B a significant factor for determining project feasibility.

The developer could choose to abandon Site A and only pursue Site B. However, we assume he/she has a strong preference for Site A and has reason to believe the proposed project will be extremely successful based on the site's highly desirable location. If a project on Site A is perceived to be less risky by lenders, the developer may be able to increase the LTV to 75 percent. We assume this will be possible although, in exchange for an increased loan amount and greater exposure, the bank will also raise the interest rate to 6.0 percent.

Using these new assumptions for Site A and keeping the original financing assumptions for Site B, we can compare the loan amounts and equity requirements as follows with calculations as shown in Fig. 4.15.

Site A Financing Assumptions

LTV: 75%

Interest rate: 6.0%

Term: 30 years

Site B Financing Assumptions

LTV: 70%

Interest rate: 5.5%

Term: 30 years

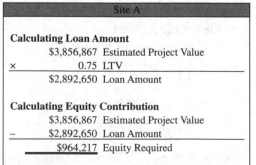

Site A	
Calculating Loan Amount	
	$3,856,867 Estimated Project Value
×	0.75 LTV
	$2,892,650 Loan Amount
Calculating Equity Contribution	
	$3,856,867 Estimated Project Value
–	$2,892,650 Loan Amount
	$964,217 Equity Required

Site B	
Calculating Loan Amount	
	$3,085,500 Estimated Project Value
×	0.7 LTV
	$2,159,850 Loan Amount
Calculating Equity Contribution	
	$3,085,500 Estimated Project Value
–	$2,159,850 Loan Amount
	$925,650 Equity Required

FIGURE 4.15 Revised equity calculations for Site A and Site B.

The developer is now confident the necessary equity can be raised for either project. However, before approaching equity investors, the developer must determine whether or not each project's NOI is sufficient to cover the loan payments by a large enough margin to satisfy the lender. Mortgage payments on commercial loans are referred to as the *debt service*, or the payments needed to "serve" the outstanding debt. Developers often use financial calculators or financial pro formas to calculate debt service, but it can be done using the following equation:

$$DS = LV\left[\frac{r(1+r)^n}{(1+r)^n - 1}\right]$$

where DS = debt service

LV = loan value

r = interest rate

n = total number of periods

To calculate the annual debt service on this loan we solve the equation as follows for each site:

Site A

$$DS = \$2,892,650\left[\frac{0.06(1+0.06)^{30}}{(1+0.06)^{30} - 1}\right]$$

$$DS = \$210,148$$

Site B

$$DS = \$2,159,850\left[\frac{0.055(1+0.055)^{30}}{(1+0.055)^{30} - 1}\right]$$

$$DS = \$148,609$$

Note that to calculate a monthly payment, the values of both the interest rate and the number of periods must be adjusted accordingly. This is done by dividing r by 12 months and by multiplying the

loan term, t, (in years) by 12 months to get n. A monthly debt service can either be compared to a corresponding monthly NOI or multiplied by 12 and compared to annual NOI. The equation for calculating monthly debt service is as follows:

$$DS = LV\left[\frac{\frac{r}{12}\left(1+\frac{r}{12}\right)^{t\times12}}{\left(1+\frac{r}{12}\right)^{t\times12} - 1}\right]$$

If the NOI is insufficient to pay, or *cover*, the debt service, then the project is deemed not viable. The developer may be able to assume a reduced loan amount, which would lower the debt service, but only if he/she can compensate by raising additional equity or reducing project costs. In this scenario, the NOI for both sites is greater than the respective debt service. Subtracting the debt service from NOI yields the project's annual before tax cash flow, as shown in Fig. 4.16.

Although the NOI covers debt service in both cases, the margin by which it does so is not the same for both sites. A *debt coverage ratio* (DCR) measures the degree to which NOI can cover the developer's loan obligation, calculated simply as

$$\frac{NOI}{DS} = DCR$$

A DCR of 1 means that the NOI is exactly equal to the debt service or, in other words, that the developer earns only enough income to pay the lender. DCRs greater than 1 mean that NOI is greater than the debt service obligation. The larger the DCR, the greater the "strength" of NOI to cover the debt service.

Lenders are naturally risk averse and want assurances that the debt service will still be paid even if NOI is threatened by unanticipated circumstances. In any given year, costs could be unexpectedly high

Site A	
Calculating Projected Annual Cash Flow	
	$231,412 Net Operating Income (NOI)
−	$210,148 Annual Debt Service
	$21,264 Before Tax Cash Flow (BTCF)

Site B	
Calculating Projected Annual Cash Flow	
	$185,130 Net Operating Income (NOI)
−	$148,609 Annual Debt Service
	$36,521 Before Tax Cash Flow (BTCF)

Figure 4.16 BTCF calculations for Site A and Site B.

or rental revenue could fall, perhaps due to the unexpected departure of a tenant. Either scenario causes a reduction of NOI and could potentially jeopardize the developer's ability to make the loan payment, which remains unchanged regardless of how operations impact NOI. Therefore, as a condition of permanent financing, lenders impose certain DCR requirements. DCR requirements can be dependent on the lender's risk tolerance, the strength of developer in terms of financials and past performance, the specifics of a particular project, or concerns about overall market conditions.

Assume the required DCR for a loan on either Site A or Site B is 1.2, as follows:

Site A Financing Assumptions	Site B Financing Assumptions
LTV: 75%	LTV: 70%
Interest rate: 6.0%	Interest rate: 5.5%
Term: 30 years	Term: 30 years
DCR: 1.2	DCR: 1.2

The DCR calculations for both sites are shown in Fig. 4.17.

On the basis of the DCR requirement, only Site B is viable. Even though Site A may still be the developer's preferred site in terms of the location and project specifics, the lender will not approve a loan and no debt financing is available for the project. Thus, Site A is not viable. The developer must decide if he/she wishes to pursue Site B and if the remaining cash flow after debt service is sufficient to compensate equity investors.

Note that during the site search stage, the developer has no rights or control over any site. Landowners can raise their asking price, sell a property to another developer, or decide not to sell at all. Once a developer has completed basic "back of the envelope" calculations and decided a particular site is worth pursuing, he/she will need to gain control of the site (Milestone 3) in order to make further evaluations of both the land and viability of the proposed project. The following sections will describe some of the contracts involved in achieving this development milestone.

4.3. CONTROLLING THE SITE

Once a developer finds a site that appears viable based on preliminarily consideration, it is necessary to perform a more thorough analysis. Development is an inherently risky endeavor and there are many factors that might adversely affect the developer's plans and investment in the property, such as title issues, boundary/survey issues, hazardous waste or contamination, land use approvals, availability of utilities, availability of funding, site access, and physical characteristics of the property itself. Studying each of these and other factors requires the developer to spend significant time and money; however, in the absence of a contract with the seller, the developer has no guarantee that the site will not be sold to another party. Therefore, in order to continue the development process and engaging in further evaluation, the developer must achieve Milestone 3: controlling the site. The developer is said to "control" a site once a legally binding contract is in place and the developer has secured an exclusive right to study and subsequently purchase the property. Two documents are typically required for a developer to gain control of a site: (1) a conceptual agreement on terms via a letter of intent and (2) a legally binding contract that has been *executed*, or signed, by the parties involved. The following sections will describe these documents in greater detail.

4.3.1. Letter of Intent

Once the developer has identified a desirable site, he/she must move toward controlling it.

Site A		
Calculating Debt Coverage Ratio		
	$231,412	Net Operating Income (NOI)
÷	$210,148	Annual Debt Service
	1.10	Debt Coverage Ratio (DCR)

Site B		
Calculating Debt Coverage Ratio		
	$185,130	Net Operating Income (NOI)
÷	$148,609	Annual Debt Service
	1.25	Debt Coverage Ratio (DCR)

FIGURE 4.17   DCR calculations for Site A and Site B.

This requires agreeing on the terms of an acquisition with the landowner/seller. Although this can happen informally with the developer and landowner meeting directly to discuss terms, it is more commonly achieved through a formal process facilitated by brokers. Both the landowner/seller and developer/buyer will likely have hired a broker to represent and assist them. Indeed, in some instances the developer and landowner may never meet and the transaction may be negotiated entirely through broker representatives with input from their respective clients. This is often the case if either the developer or landowner is not locally based or is a large, commercial firm rather than an individual.

After preliminary discussions about a site, one of the brokers will create an initial written proposal for the transaction. The proposal document will specify basic terms, such as the size and location of the property as well as a proposed purchase price. It will also include critical or unique terms, such as the length and purpose of a nonstandard study period; the obligation of the landowner to deliver vacant possession if the site is occupied; or the responsibility for the removal of existing structures, if applicable. Proposals may be exchanged several times as each party refines its position and counters with a revised offer. It is not uncommon for this process to take several months, particularly for complex transactions. During this time the brokers will act as intermediaries and represent their client's interests while working to bring the parties to agreement if possible. In complex negotiations, the brokers may facilitate in-person meetings, when possible, to enable a more direct exchange. In especially complex negotiations, the parties' attorneys may also become involved in negotiations, although more typically, the business teams work out the terms of the deal and then solicit their respective attorneys to prepare or review a draft offer letter or draft contract as applicable.

If agreement on terms is not possible, the developer will have to continue searching for another site. However, if terms are reached that are acceptable to both parties, one of the brokers will convert the proposal document into a letter of intent. A *letter of intent* (LOI) is a nonbinding document that formally captures the various specific deal terms that have been agreed to by both parties. Note that in some markets the same document is known as a memorandum of understanding (MOU), but LOI will be used herein. The length of an LOI depends on the complexity of the deal and level of specificity required. They can be as short as a single page but are often several pages in length. LOIs themselves may be subject to further negotiation if they contain specific language that one or both parties want incorporated directly into a contract clause(s). However, LOIs include explicit language stating that they are not legally binding; either party can deviate from the terms contained in an LOI, but doing so is considered an act of bad faith and may cause the deal to be terminated. Although they are nonbinding as to the obligation to purchase or sell the property, LOIs include a signature line so that each party can indicate acknowledgment of the terms, which will then be used to create a binding contract. LOIs serve as an important mechanism to ensure agreement on business terms before the parties engage attorneys and begin to incur legal fees for contract negotiations.

However, because an LOI creates the basis for the more detailed legally binding contract, the developer may have his/her attorney review the unexecuted LOI to ensure its terms are correctly incorporated into the resulting contract. Developers will often execute an LOI without attorney review and subsequently retain an attorney to draft the contract. This can add difficulty to the process if, for example, the attorney identifies a provision of the LOI that was not drafted with the developer's interest in mind. Although nonbinding, issues that were not properly addressed in the LOI can become sticking points if they are changed at the contract-drafting stage. Therefore, for substantial projects, it is advisable to have an attorney review the LOI before the developer signs it.

It should be noted that site acquisition is not the only time that developers use LOIs. The developer will also hire a commercial broker to sell or lease space in the proposed project, often before construction has begun. LOIs will be used to facilitate the resulting transactions. Commercial brokers often specialize in either sales or leasing and can also specialize by type of product: office, retail, and industrial. They will often be part of a company of brokers that includes specialists in all of these areas so they can assist a developer with whatever type of project he/she may have. For example, if the developer is planning an office project, he/she will hire a broker specializing in Office Landlord Representation. In this case, the developer is known as the *landlord* and the leasing party is the *tenant*. An LOI for office space in the proposed project will contain all necessary business terms, including rental rate, rental abatement (if any), tenant improvement allowance, annual escalations to rent, operating expense base year, any rights to renew and expand or reduce suite size, signage rights, parking rights, and building amenities. The LOI will serve as the basis for a commercial office lease document that enables the tenant to occupy the newly built space and the developer to receive rental income on the property.

4.3.2. Contracts

A contract is a legally binding agreement between two or more parties. Under U.S. law, a contract must satisfy four conditions: (1) be of legal purpose, (2) be made between competent, legal adults or entities, (3) include the offer and acceptance of some action, good, or service, and (4) envision the payment of something of value, known as *consideration*, in exchange for performance. Contracts for the purchase of real estate must satisfy a fifth requirement: they must be made in writing in order to be legally binding on the parties. The next sections will examine contracts in the context of a development project.

Contracts Overview. The developer's attorney plays a critical role in the negotiation of the different contracts necessary for a development project and actively assists the developer in identifying and managing risk in the acquisition of land for development. Generally, there is little involvement from other members of the development team during the contract negotiation stage; yet elements of certain contracts can impact different team members during the development process. For example, a contract may specify the length of time available for certain studies or in which to obtain approvals.

The development team will be held to these time periods, as negotiated in the contract, or else the developer may face significant financial consequences and could even default on the contract and lose control of the site.

Civil engineering firms, lenders, and other team members that have close and on-going professional relationships with a particular developer may be consulted for input with respect to contract terms that impact their discipline. Even if they have not yet been hired for a project, it may behoove civil engineering firms and other team members to volunteer input with respect to site-related challenges the developer should consider addressing in the contract. This can serve both a business development strategy and also as a preemptive effort to ensure sufficient working time should the project come to fruition. However, any such input will inherently be based on imperfect information given that, prior to contract signing, neither the developer nor any potential team member will have access to verified site information. Unfortunately, if the developer is unsure which firms he/she intends to work with on a project or chooses not to solicit input, it is unlikely that professionals from the development team disciplines will have the opportunity to offer suggestions.

Types of Contracts. There are several types of contracts used to facilitate development projects. Some are related to the acquisition of the site while others formalize the relationship between the developer and his/her investors or partners in the deal. This section will introduce some of the most common contract types and describe their uses and common clauses.

Purchase and Sale Agreement. A *purchase and sale agreement* (PSA) is a contract that legally obligates the developer to purchase a site subject to certain conditions being satisfied. These conditions may be tied to predetermined dates and can be related to site conditions; external factors, such as obtaining financing; or may be related to certain seller obligations, such as delivering a vacant site free from occupants. Some conditions may even be dependent upon the completion of other conditions, which are referred to as *conditions precedent*.

Upon signing, or *ratifying*, the contract, the developer will be obligated to make a substantial payment to the seller in the form of a deposit. The deposit may or may not be recoverable if the acquisition ultimately does not occur.

The terms memorialized in the PSA constitute the entire agreement between the developer and the seller of the land. Any circumstance that arises during the course of the transaction, foreseen or unforeseen, will be handled as directed by the contract (or else result in a civil suit), so it is important to the developer that the contract be comprehensive. As a result, PSA contracts can be long and complex due to the vast range of conditions they envision. They are often even more detailed for large, complicated deals such as those requiring zoning changes, phased assemblages, or the removal of existing tenants. It is essential that the developer and his/her attorney are diligent in negotiating terms to mitigate the risk of unforeseen or undesirable consequences. In this sense, the PSA is a tool for managing risk as much as it is a vehicle for the purchase of land or property and the developer will dedicate significant time and financial resources to work with his/her attorney in order to negotiate the PSA.

It is important to note that because of the time and cost involved in negotiating a PSA, the developer and seller will first attempt to agree informally on important terms, such as price. This is achieved through the use of an abbreviated, deal summary document, usually an LOI as previously described. While LOIs generally provide that they are not binding and that the specific terms will be documented in a subsequent written agreement, they serve an important purpose by helping the developer and seller confirm that they are in substantial agreement on key terms before incurring more substantial attorney fees to negotiate a formal contract.

Standard Clauses. As previously stated, the PSA represents the entire agreement between the developer and the seller. As such, the contract will not only establish a purchase price, but will also clearly define all terms, provide detail with respect to related factors, and specify the conditions of the sale. Note that the seller and the developer (the buyer in this context) have fundamentally different perspectives with respect to negotiating contract terms. The seller wants to dispose of the property as quickly as possible without further worry or obligation; meanwhile, the developer/buyer wants assurances about site conditions and time to address concerns about the site's viability, assemble a team, obtain investors, secure debt financing, order technical studies, and begin the design process. As a result of the tension between the seller and developer perspectives, contract clauses are not uniform and must be negotiated for each transaction to satisfy the individual parties involved.[2] Some of the most common PSA components and clauses are introduced next.

Define the Parties. The PSA will clearly define the parties to the contract as well as describe how the developer will hold title to the property once acquired (i.e., partnership, corporation, LLC, etc.), which has both tax consequences and liability protections.

Costs: Purchase Price, Deposit, and Other Cost Allocation. The purchase price and any payment schedule (if applicable) will be clearly defined. Typically, a purchase price consists of a cash payment made by the developer (usually funded by equity partners) and a promissory note secured by a deed of trust or mortgage. The deposit required by the seller will also be specified along with specific conditions under which it is refundable or will be forfeited to the seller. Retaining the deposit is often the seller's only remedy if the developer defaults on the contract, so sellers naturally want the posted deposit to be as large as possible and available for capture at the earliest possible date. Note that the point at which the seller *can* retain the deposit is not necessarily the same as the point at which the seller does so. Rather, once the deposit is "at risk," it exists as a remedy if the developer subsequently decides not to purchase the site or defaults. If the developer proceeds with the deal, the contract typically specifies that the deposit be counted toward the purchase price.

The PSA will also describe the allocation of transaction costs and costs associated with rollback taxes, special assessments, proffers, and broker commissions. These are usually allocated based on existing custom in the market, but can be negotiated as with any other deal term.

Financing. If the development requires a loan, as most large projects do, the developer will require a clause allowing a *financing contingency*, which provides the developer a period of time within which to obtain financing for the acquisition of the site and the subsequent construction. If the developer cannot obtain financing within the allotted time, the contingency is not satisfied and the developer is relieved of his/her obligation to complete the transaction. In such cases, the developer's deposit is generally returned. Financing contingencies ensure that the developer does not become legally required to purchase a site when he/she has no financial means of doing so, which could lead to bankruptcy or other severe consequences. However, sellers are not necessary willing to offer open-ended financing contingencies in which a developer could purposely fail to obtain financing as a strategy for not completing the purchase. Thus, while a seller may grant a financing contingency, the contingency will be subject to certain terms and require the developer demonstrate best efforts to obtain financing.

Representations and Warranties. Given the amount of time and money involved in completing a commercial transaction, both the seller and developer want assurances pertaining to basic yet materially relevant facts. This is achieved by the incorporation of representations and warranties into the contract. These are formal statements that the parties can be held to legally. For example, the developer will need to assure the seller that he/she has the authority to enter into the agreement on behalf of his/her company and/or investors. Further, the developer may be required to assure the seller that he/she is not facing bankruptcy and, therefore, is able to pursue the acquisition in good faith. Equally, the developer will want the seller to make certain representations about the site and provide warranties with respect to certain conditions. These representations and warranties from the seller help minimize the developer's risk because, if they are found to be incorrect, the liability for the error will fall to the seller. Sellers, on the other hand, seek to minimize their risk by making the sale "as is," ideally providing no representations and warranties about the property, or at least limiting the representations and warranties as much as can be negotiated. Examples of seller representations and warranties to be negotiated include:

- In the event that the entity selling the property is a corporate organization, that the person negotiating with the developer has the authority to enter into the transaction as well as the authority to execute relevant documents.

- The seller is the fee simple owner of the entire property.

- The property is free and clear of all liens and encumbrances.

- The property is not subject to any unrecorded restrictive covenant or equitable servitude of any kind that would in any way limit the free choice of the developer with respect to the nature or location of the contemplated development.

- There are no hazardous substances on the property and no historical disposal of hazardous substances that would require remediation under environmental laws.

- There are no graves on the property.

- There are no lawsuits threatened or pending.

- The seller is not bankrupt or insolvent (or about to be).

- The seller has not given any other party or person the right to use, or an interest in, the property.

- There are no current notices from the government of special assessments.

- Execution of the agreement will not conflict with or result in a breach of any agreement to which the seller is a party (including loan documents).

- There are no mechanics liens.

- The seller will provide a closing affidavit.

Negative Covenants. In a PSA, a *negative covenant* is an agreement not to do something that would adversely impact the sale or property. This type of clause generally requires the seller to maintain the property in its present order and condition during the contract period through to closing. This provides the developer with assurance that he/she will not discover unexpected conditions upon or after completing the purchase. Examples of common negative covenants include the seller's obligation:

- Not to alter, amend, or take any action which could result in the revocation or invalidation of any permits for the property.

- Not to place, or consent to the placement by others, of any easements, covenants, restrictions, or encumbrances without obtaining the developer's prior written consent.

- Not to enter into any other contracts or negotiations for the sale of the property while the current PSA is pending.

Feasibility Period. Perhaps the most important PSA clause from the perspective of civil engineers and other development team members is the clause establishing a feasibility period. The *feasibility period*, also referred to as a "study period," "inspection period," or "due diligence period," is a period of time in which the developer and the development team will be able to access the site and conduct physical inspections. During this time, the developer will also conduct legal and financial studies related to the project. Developers want as much time as possible during the study period to coordinate and hire experts, source funding, and evaluate the property. However, sellers typically bear the risk that a property may suffer damage between the date the contract is signed and closing, or that another offer will be lost. Thus, sellers generally want the shortest possible time periods for contingencies and studies so that closing can occur as quickly as possible.

The amount of time for the feasibility period negotiated in the contract is critical because once the period expires, the developer will have no more time to evaluate the site and will need to make a final decision about purchasing the property. Although it is possible for the developer to

attempt and negotiate an extension with the seller, this may only be possible if certain conditions were discovered that both parties agree materially change the understanding and require further study. However, the seller is under no obligation to grant an extension and may charge an additional, nonrefundable deposit in exchange for doing so.

While the study period is negotiable, 90 days is a common length of time for a study period on raw land for commercial projects. During the study period, the developer retains experts/consultants in the various risk areas (e.g., wetlands/environmental consultant), who conduct inspections and provide analyses and reports. Typically, the developer may choose, at any time prior to the expiration of the study period, whether or not to purchase the property. The decision to discontinue pursuit of a site as permitted by the PSA agreement is sometimes referred to as "kicking" or "kicking out" the contract. If the developer decides not to buy the site during the allowed time, he/she has no further obligation and any deposit paid to the seller is usually returned in full. Note that the deposit often becomes nonrefundable if the developer elects to continue with the purchase after the expiration of the study period. This is often referred to as "going hard on the contract" and means that the deposit will be retained by the seller if the developer does not subsequently purchase the property for any reason.

Examples of the types of studies undertaken during the feasibility period, many of which are discussed in detail in Chap. 5, include the following:

- Title review

- Boundary survey

- Environmental assessment

- Site condition assessment

- Zoning review

- Market survey

- Architectural or structural engineering report (if applicable, as may be the case in a redevelopment scenario)

- Review of existing leases (if applicable, as may be the case in a redevelopment scenario)

Contingency for Rezoning or Other Land Use Approvals. In some instances, the developer already knows that the subject site does not have the correct zoning in place for the intended project. In these cases, the developer will attempt to negotiate a contingency for rezoning or approvals into the PSA such that he/she does not have to complete the purchase unless the necessary permissions are granted. Such a provision makes the developer's obligation to purchase the property subject to his/her ability to obtain all necessary land use approvals and permits (e.g., subdivision, site plan, and zoning approvals, or approval by architectural review board). Effectively, this makes zoning approvals a condition precedent to each party's obligation to close. If the necessary approvals are not obtained, the developer will have incurred out-of-pocket expenses, but gets the deposit refunded and is not liable for failing to close. Note that such a clause favors the developer because the approvals could take a significant length of time to pursue and may not be granted, in which case the seller does not realize a sale and may have forgone other offers in the meantime. Thus, even if the developer is willing to offer a set amount of time within which to obtain approvals, the seller still might not be willing to reserve the land for the time necessary to obtain all applicable land use approvals or the seller may require additional payments as consideration for keeping the property off the market.

Note that it may not be worth negotiating a rezoning contingency in all cases. The significant cost involved in creating and submitting plans, with no guarantee of approval, may mean that it is not worth attempting to rezone in the time period specified in the PSA, if at all. Some developers have no interest in taking a project through an approval process and it may make more sense to evaluate the feasibility a project permitted by right under the existing zoning and negotiate a lower purchase price for the property (if possible). Similarly, when contemplating a site, the developer might be searching for opportunity and not yet know exactly what type of project he/she will pursue. In this case, the developer's uncertainty about the project means he/she does not know what rezoning or approvals might be needed. This makes it difficult to justify the time and cost to pursue approvals that that might later prove undesirable.

Option Contracts. While many developers use PSAs to acquire property, other types of contracts can also be used. Another common contracting vehicle is an option contract. An *option contract*, as the name implies, gives the developer the unilateral right, or *option*, to purchase a particular site during a specified period of time if he/she chooses to do so. Essentially, the developer is buying the right to temporarily reserve a site and prevent someone else from buying it. If he/she is willing to spend the money, a developer can option multiple sites simultaneously. This strategy is often employed in the context of assemblages where the developer may need to control several properties at once yet be unwilling to complete any purchase until the assemblage has proven viable. While the site is under option, the developer has time to perform studies and/or seek approvals. If the developer wishes to exercise the right to purchase the site at any time prior to the option expiration, he/she is said to *exercise the option*.

Standard Clauses. Option contracts may be less comprehensive than a PSA and only include general terms under which the developer can purchase the property, such as including a predetermined price, with the full purchase contract details to be negotiated upon exercise of the option. However, it is also possible for an option contract to include a fully pre-negotiated PSA as an addendum that will come into force once the option has been exercised.

Regardless of the level of detail with respect to a potential future purchase, virtually all option contracts will obligate the developer to make a payment to the seller in the form of an option fee. This fee is usually not be recoverable if the developer decides not to purchase the property and the option is not exercised. The fee may or may not be applied to the purchase price in the event that the developer pursues a purchase of the site. The cost of the option can be any amount agreed to by the parties: it may be negotiated to cover seller's opportunity cost of not selling the property to other interested parties; be set as a premium above the seller's carrying costs (i.e., the amounts the seller is paying in interest on an existing mortgage and property taxes); or it may be tied to the length of the option, with a seller requiring a larger fee in exchange for a longer option period. Option contracts can contain renewal clauses allowing the developer to extend the option period, usually in exchange for additional fees.

Public-Private Partnership Agreements.

Public-private partnerships, also called P3s, are projects in which a public sector government agency works in partnership with a private sector developer in order to complete a public-sector project more cost effectively and efficiently than the government could achieve in isolation. P3s are often used for infrastructure and utility projects. In all P3 arrangements, both the private sector developer and the public sector agency make important contributions to the project. The private sector developer typically contributes funding (from equity investors and lenders), planning and construction services, and may even manage or operate the completed asset. A typical public sector contribution often comes in the form of government-owned land that will be used for the project. The P3 agreement should contain all necessary terms of the partnership. P3s will be discussed in greater detail in Chap. 7.

CHAPTER CONCLUSION

Selecting and controlling a site are two critical development milestones that can often be time consuming and difficult to achieve. A developer may have to evaluate several sites over the course of many months before finding one that represents a viable opportunity. Once a suitable site has been identified and a contract successfully executed, the developer must evaluate the site in greater detail in order to ensure project feasibility and make an informed purchase decision. This involves several members of the development team. The next chapter will discuss the different elements of due diligence (Milestone 4) and site acquisition (Milestone 5).

REFERENCES

1. Federal Reserve website, "About the Fed," available at https://www.federalrescrve.gov/aboutthefed/structure-federal-reserve-system.htm (accessed January 6, 2018).
2. M. Jay Yurow, "Commercial Real Estate Transactions," 14 The Compleat Lawyer 42, Spring 1997.

CHAPTER 5

DUE DILIGENCE AND SITE ACQUISITION

Once a developer controls a site that appears viable based on preliminary considerations, it is necessary to perform a more thorough analysis. Development is an inherently risky endeavor and there are many factors that might adversely affect the developer's plans and/or investment in the property such as title issues, boundary/survey issues, hazardous waste or contamination, availability of utilities, availability of funding, site access, entitlements (zoning and permit requirements/processes), and physical characteristics of the property itself. Studying each of these and other factors requires the developer to spend significant time and money during the contractually allowed study period as part of a process called *due diligence*. Upon completion of due diligence studies, the developer will have gathered and verified sufficient information to make an informed decision about whether to pursue the proposed project or deem it nonviable. If the developer chooses to proceed, information gathered during due diligence will be used in subsequent phases of the development process from preparing a submission for a loan to submitting the site plan. Because it is such a critical part of any project, completing due diligence (Milestone 4) is a critical milestone in the development process.

Due diligence is especially important in real estate transactions because U.S. law does not obligate sellers to represent or warrant the condition of a property. This is succinctly captured by a Latin phrase that is often used in the real estate industry: caveat emptor, or "let the buyer beware." In residential transactions, some state laws require the seller to make certain disclosures, such as regards known defects; however, with respect to most aspects of a commercial real estate purchase, courts still apply the doctrine of caveat emptor. Nevertheless, a commercial purchaser typically will have a more substantial list of issues to investigate than a residential homebuyer and, therefore, study periods for commercial transactions tend to be longer than for residential purchases. Given that it is the developer's (buyer's) duty to conduct due diligence in the study period, the seller is motivated to make as few representations and warranties as possible in the contract, and will generally seek to avoid giving any assurances to the developer (buyer) that the property is fit for any particular purpose. Therefore, unless the seller makes specific representations and warranties that will survive the purchase contract, the developer bears the risk of future problems arising when acquiring land or commercial properties. Therefore, again, successful completion of due diligence is a critical development milestone.

This chapter is dedicated to three main forms of due diligence: legal, technical (site based), and financial. It will also discuss the associated development milestone of site acquisition (Milestone 5), or "closing," at the end of the due diligence period. It behooves all development team members to not only be familiar with elements of technical due diligence, but to also be aware of the developer's legal and financial considerations during this stage. Note that the purpose of this chapter is not to enable team members to undertake the various forms of due diligence activities outside their area of expertise, but rather to raise awareness of the many different issues the developer must consider during this critical stage of the development process.

5.1. LEGAL CONSIDERATIONS

The results of site-based and technical due diligence can often require legal guidance. This may arise in relation to environmental issues, title defects, zoning restrictions, or other matters. The developer's attorney will help review pertinent information and advise the developer with respect to the implications for the site acquisition and proposed project. The attorney's role in evaluating the results of technical and site-based due diligence will be discussed throughout this chapter as applicable. In addition to legal due diligence associated with the site itself, the developer must consider several important investment and operational issues related to the project that are facilitated by contracts and usually require the assistance of the developer's attorney. The following sections will describe some of these considerations, such as forming the legal entity used to acquire the site.

Development projects are complex and involve a myriad of contractual, regulatory, and other legal issues. Among the attorney's primary roles are assisting the developer in navigating these issues as well as identifying and managing risk in the acquisition of land for development. The following sections outline important legal considerations that need to be addressed. Note, however, that these may not always align with the contractual due diligence period.

Establishing a Limited Liability Corporation.
Commercial developers rarely own real estate personally. Doing so creates a liability exposure that can put their personal assets at risk for claims in lawsuits, loan defaults, and operating deficiencies. Rather, the developer's attorney will typically create a legal entity that takes title to the subject property. When the entity is created solely for the purpose of holding a particular piece of real property (i.e., a single asset), it is referred to as a *single purpose entity* (SPE). The most common type of legal entity used is a limited liability corporation (LLC); however, other legal entities such as corporations, partnerships, and trusts may also hold title to real property. LLCs are often used by real estate investors to hold real property because an LLC offers the benefits of pass-through taxation of profits and losses, limited liability protection for owners, and relative ease of administration. LLCs (and other legal entities) may perform many of the same functions that people can, including opening bank accounts, obtaining loans, paying taxes, and owning property. An LLC can purchase and develop property it its own name and receive income from the operation or eventual sale of a development project. Under an LLC structure, the developer is a member of the LLC and, if the LLC operating agreement provides, the developer can receive compensation from the LLC.

Developers may own and operate multiple LLCs. It is not uncommon for a developer to set up a new LLC for each development project. This helps to ensure that the assets of each individual LLC are protected from claims arising from properties owned by other LLCs. When pursuing a new project, the developer may purchase a property and then transfer it to an LLC or may use the LLC to make the initial purchase directly. In either case, and in the event a different type of entity is used (for tax and other considerations), the developer's holding strategy impacts site acquisition and development decisions.

Joint Venture Agreements.
In the context of real estate, a *joint venture* (JV) is an agreement between at least two parties for the purpose of facilitating the acquisition or development of a property. While the terms of a joint venture agreement

are typically formalized by a written agreement, the parties to a joint venture agreement may or may not be partners or members of an LLC or other legal entity (discussed previously). Whichever legal entity structure is utilized, a typical JV includes one party as the operating member and the other party (or parties) as the capital member. The *capital members* provide a significant capital contribution in the form of equity that is used to fund the project often in conjunction with additional debt financing. The *operating member*, who may also be the *managing member*, is responsible for directing the development and business activities of the venture. Usually, the developer acts as the operating member, which allows him/her to retain significant control of the project and often limits the amount of capital that he/she must personally contribute. However, the operating member also may be the one who is personally liable to repay any financing liabilities if the project fails to meet its obligations to the lender.

If the parties' joint venture agreement is embodied in an LLC structure, for example, then the parties will enter into an operating agreement (which sometimes may be referred to as a joint venture agreement), which formalizes their roles, relationships, rights, and duties. Given that most large-scale development projects can only happen with the contribution of multiple parties, JV agreements are common in the industry, and typically the parties to a joint venture will negotiate the terms of their relationship before or at the same time the developer is negotiating a purchase and sale agreement or option contract.

Standard Clauses. The JV agreement (or, for an LLC, the operating agreement) will outline key components of the relationship between the members, such as how profits are distributed; how (and by whom) the development project will be managed; the manager's control over certain decisions versus member control; what decisions require a member vote and what percentage vote is necessary to carry which decisions; and the transfer of membership interests, which is usually prohibited without the consent of other members and/or under certain conditions. Some common JV agreement clauses are described next.

Capital Contributions. The operating agreement will establish the capital contribution obligations of the parties and whether members can be required to make additional capital contributions as well as consequences for failure to meet such obligations. The type of asset being purchased may factor in to terms regarding additional capital. For example, in the case of investing in an existing building, the capital member may not agree to make additional capital contributions, whereas, the parties may agree to fund their share of budgeted development costs for development projects. In the context of development JVs, the operating agreement should clearly address how the parties will share not only in budgeted costs but also cost overruns and unforeseen costs. The agreement may require mandatory additional capital contributions for certain nondiscretionary expenses, such as real estate taxes, insurance, and legal compliance. When the managing member formally requires additional capital contributions, it is referred to as a *capital call.* Typical remedies for a member's failure to fund a mandatory capital call include the following:

- Use of a member loan allowing one party to fund another party's share. The loaning member will be entitled to return with interest before any future returns are distributed to the nonfunding party.

- Dilutions, which are a form of penalty that decreases the nonfunding member's interest in the joint venture and increases the funding members' interest.

- Default clauses that can trigger a buy-out or other remedies.

Profit Distribution. In the context of real estate development, joint ventures are profit-seeking endeavors. Naturally, therefore, JV agreements include language directing how the project's profits are distributed between the members. Capital members expect not only the return of their original equity investment, but also the payment of additional funds. Capital members typically receive a priority payment, known as a *preferred return*, until the entirety of the original equity investment has been repaid in full. Afterward, capital members receive ongoing payments at a pre-agreed rate of return. The operating member (usually the developer) also expects to receive returns and, in exchange for creating the development opportunity, is often entitled to receive a percentage of the distributions that is greater than his/her actual percentage interest in the joint venture. This additional financial incentive for the operating member (developer) is often referred to as a *promote*.

If additional profit is realized after both the operating and capital members have received their guaranteed payments, a profit-splitting mechanism is used to distribute any remaining funds. This distribution mechanism, known as a *waterfall* because of the cascading distribution splits, describes which party receives what portion of any net operating proceeds and/or capital proceeds from the asset. However, if profits are insufficient, the operating member may forgo payment because of the priority given to the capital members. If the project fails to generate enough profit to make payments to capital members (usually in a particular year of operations), the amount owed to the capital members will accrue and receive priority for payment from future profits.

Management Structures.[1]   The JV agreement will specify the management structure as well as which decisions can be made at the manager's sole discretion and which require some form of member approval (majority, supermajority, or unanimous). Examples of typical JV management structures in an LLC include the following:

- Manager/member-managed with one manager in control
- Manager/member-managed with joint control
- Manager/member-managed with one member in control subject to certain major decision rights
- Board of managers

The managing member in a joint venture is often entitled to collect management fees in exchange for services. These may be specified in the JV agreement but can also be captured in a separate management agreement. Typical examples of fees include an:

- Acquisition fee
- Financing fee
- Asset management fee
- Property management fee
- Leasing fee
- Development fee
- Disposition fee (for overseeing the sale of the completed asset)

In the capacity described by the agreement, the managing member is responsible for making key business decisions pertaining to the JV's real estate investment. Typical examples of major decisions are:

- Acquiring new real assets (including land for new development projects)
- Material capital expenditures
- Making profit distributions
- Sale or refinance of the property
- Filing for bankruptcy protection
- Deadlock resolution mechanisms
- Events triggering buy-outs between members
- Dissolving the company
- Right of first refusal

Rights Related to the Sale of Member Interest. Circumstances, whether external or related to the investment, may motivate one or more JV members to divest of their interest in the project. The JV agreement provides various mechanisms to direct the potential buy-out of member interests. For example, *right of first refusal* provisions require that if a member receives an offer from a third party for his/her membership interest, he/she must first give the other members the option to buy-out his/her membership interest. If a member does not have a third-party offer, a "put option" allows that member to require the remaining members to buy out his/her interest. A "call option" gives a member the option, but not the obligation, to purchase the interest of the other member(s) on certain terms specified in the agreement. *Tag-along rights* give a member the right to voluntarily "tag along" with a selling member in order to (1) benefit from a desirable sale price and (2) prevent the member from being forced to remain in the JV with a new, unknown partner after an outside party buys out other members' interests.

5.2. SITE-BASED DUE DILIGENCE

In addition to legal due diligence, studies of the subject site itself must also be undertaken during the study period in order for the developer to fully complete his/her due diligence and achieve Milestone 4. For purposes of this book, "site-based due diligence" refers to studies physically conducted on or related to the subject site. These generally fall under two categories: (1) technical surveys and studies, and (2) considerations related to site design. The following sections will discuss key site-based due diligence activities and implications for the developer.

5.2.1. Technical

The following sections outline some of the many technical tasks undertaken during the due diligence period. These include title surveys, boundary and topographic surveys, geotechnical reports, and environmental reviews. The information presented here is meant as an overview in the context of the development process; additional, technical detail on these and other forms of technical due diligence can be found in the *Land Development Handbook*, Fourth Edition.

Title. As briefly described in the previous chapter, the most complete form of private ownership of real property available in the United States is known as *fee simple absolute*. The actual process of securing this ownership is achieved by the registration of a deed of title, also sometimes referred to as a title deed. Although used in combination, a "deed" and a "title" are two separate components of property ownership. *Title* is intangible; it refers to the status of ownership of a property. A *deed* is a written document that serves as the legal

mechanism to show evidence of the transfer of real estate ownership. After a transfer of property occurs, the deed showing evidence of the transfer and the new ownership must be filed with the local jurisdiction, usually at the County Recorder's Office, in order to be formally recognized and valid. This is known as *recordation* or *registration* of the deed. Collectively, all the documented deeds for a property create a record that is often referred to as a "chain of title" where each individual deed represents a link in the historical record (or chain).

The title history for a parcel of land must be checked as part of the due diligence process before the developer acquires the property. This is required by lenders and also gives the developer confidence that there are no hidden problems with the property. The title review is usually performed by a specialized title company. A *title abstractor* searches the land records to confirm the chain of title and to see if there are any liens, easements, covenants, restrictions, or other matters affecting ownership and use of the property.

Liens are claims filed by creditors that are attached to real property in exchange for payment. A sale of property must satisfy any outstanding liens or else the sale may be prevented. In real estate, the two most common liens are a mortgage lien and a mechanic's lien. A *mortgage lien*, which will be explained in greater detail later in this chapter, is a record of a bank's claim of repayment for a loan. *Mechanic's liens* can be filed by contractors against a property if the owner fails to pay for their services.

An *easement* is a right to access or use a property by a person or entity that is not the owner of the property. In other words, it is a nonpossessory right to real estate that benefits someone other than the property owner. A common example is a utility easement, wherein a utility company runs pipes or wires across (or under) a property. However, an easement can also be granted or created in favor of private individuals or companies. Generally, a developer cannot build on an easement area, so it is important to know if these exist and, if so, where they are located on a site. In some instances, the developer may be able to negotiate with the easement holder to move the easement location, usually in the case of relocating utilities, but the cost to do so will usually be borne by the developer and can impact project costs substantially.

In the context of commercial real estate, a *covenant* is a legally binding agreement tied to the use of a particular property that is recorded on the property title. These can be put in place during the filing of an original plat or as a condition of conveyance that attaches to the title during a sale. Covenants bind all current and future owners of the property. For example, a covenant dictating that a particular parcel can only be used as a school for a 50-year period would prevent a new retail center (or any other use) from being developed on the site until after expiration of the restriction, regardless of how many times the property was sold during the 50 years. Developers need to make sure that no problematic covenants exist when acquiring a site or, if covenants do exist, determine how they will impact the proposed project. Note that when creating a new project, developers often impose their own covenants. For example, restrictive covenants are often put in place by residential developers to ensure that the vision for a residential community is preserved in the future. This may include a restrictive covenant on paint colors, architectural design, or materials that can be used for homes and other structures in the community. Examples of covenants on commercial projects include restrictions on signage or operating hours.

At the end of the title review, the title company prepares a title report, also called a *title commitment*, based on the results of the search. The report lists encumbrances, exceptions, and other matters of record. Based on the report, the title company will then issue a title insurance policy that benefits the lender, although policies can also be issued to benefit the developer. Title insurance is unlike other forms of insurance in that its purpose is not to indemnify the policyholder for a future, unanticipated loss. Rather, the premium is paid primarily to cover the cost of the title examination, which is meant to identify any title problems prior to consummation of the transaction. The title company will generally require any title deficiencies it identifies to be corrected before the close of the transaction and items that are not corrected

will remain as "exceptions," which are excluded from coverage. These exceptions may be eliminated by correcting the deficiency. For example, the title insurer will require any existing mortgage liens, judgments, and other similar encumbrances to be satisfied and released prior to closing. Or, if any of the exceptions identified in the title report are cause for concern and cannot otherwise be eliminated, the developer can mitigate that risk by obtaining specific endorsements, for a cost, to address particular concerns.

The developer and his/her attorney will evaluate the legal ramifications of any issues listed in the title report. In some cases, it may be possible to mitigate impact on the proposed project. For example, it may be possible to propose the relocation of a utility easement (and any utility lines/pipes) as part of a comprehensive rezoning package. However, such results are never guaranteed. This may mean that the site layout must be redone several times or ultimately may not be optimal. This can cause frustration to members of the design team but is often unavoidable.

Boundary and Topographic Surveys. The two most common site surveys the developer will need to pursue during the due diligence period are a boundary survey and topographic survey. These are necessary to confirm the subject site's boundaries and other existing conditions that can impact the proposed project. Both surveys are an integral part of the preliminary design process and the omission of valuable information they provide can lead to costly problems later on with the project. While these surveys provide preliminary information suitable for the due diligence process, more detailed surveys will eventually be required for final designs.

A *boundary survey* is the result of a process by which a professional surveyor identifies the property lines and easements within a tract of land. This is extremely important as all future work related to the proposed project hinges on the initial survey work. Should an error occur in the survey, it will most likely manifest itself in later stages of the development process. It is conceivable that errors in the initial survey could lead to improper location of streets and other improvements and to incorrectly placed interior boundary lines. Errors such as these can cause project delays and the need to redesign the site, which inevitably costs the developer both time and money.

When the proposed project will involve one or more loans, the lender typically requires certain assurances that the parcel boundaries stated in the contract match both the title records and the actual site area being marketed for sale. To address this need, the American Land Title Association (ALTA) and the American Congress on Surveying and Mapping (ACSM) jointly developed and adopted a set of standards for conducting surveys in conjunction with the closing of commercial loans for the acquisition of real property. The ALTA/ACSM Minimum Standards provide for a nationally defined and accepted set of survey requirements and are incorporated by surveyors into an ALTA/ACSM Land Title Survey, which will be included in the developer's loan application materials. The Minimum Standard Detail Requirements for ALTA/ACSM Land Title Surveys outline the responsibilities of the surveyor in conducting a survey that will be used by the title company and lender in conjunction with the closing of a commercial loan on real property. The standards also outline the responsibility of the developer with respect to providing complete sets of documents from the title research and as may otherwise be required by the surveyor to conduct a proper and complete survey. In addition, the survey contains a mechanism that allows the developer or lender to request additional information from the surveyor related to certain specific issues such as availability of utilities.

The ALTA/ACSM Minimum Standards call for the surveyor to resolve the boundary of the property in addition to providing comprehensive documentation of any facts on the ground or in the provided records that may be evidence of otherwise unknown or undisclosed title problems. This might include, for example, the discovery of potential encroachments and/or other uses of the subject property by others. The disclosure of these facts allows the title company, lender, and developer to weigh the associated risks and to negotiate or formulate appropriate resolutions to any issues that might have been found during the survey.

While the boundary survey is the delineation of property lines and easements, a *topographic survey* is the delineation of natural and artificial physical features. Although the topographic features of a property may have some significance to project feasibility and site analysis, their primary importance is initially realized in the early design and engineering phase. The horizontal and vertical positions of natural and man-made features play a critical role in determining how a site will be engineered to suit the intended use of the property. Consideration must also be given to the way these features relate to those lying on adjoining properties and within street rights-of-way or easements. Determining how utility services can be provided to the proposed development is among the first preliminary design tasks undertaken.

Topographic information for areas well outside of the subject site boundaries is often needed to make preliminary decisions regarding design considerations such as storm drainage patterns. When this is the case, the use of smaller-scale topographic maps, such as those available from the U.S. Geological Survey (USGS) with a scale of 1 inch = 2,000 feet or from local GIS data, is appropriate. As the design process begins to focus on more intricate details, larger-scale maps are needed to ensure that site engineering is as balanced as possible with regard to earthmoving and the use of utilities. The size and relative detail of a site often determine the surveying methods employed. For large areas, aerial photography and photogrammetry are economical means of providing detailed maps. These surveys have become increasingly more accurate with improved technology, but always need to be verified. Moreover, there may be site features that will need additional detail such as areas obscured by tree cover. Field methods employing electronic survey systems may be used on other sites to create quality maps in a timely fashion.

Geotechnical Reports. A *geotechnical report* communicates pertinent information related to existing site and subsurface conditions, which are then incorporated into recommendations regarding site and building design based on the geotechnical engineer's findings. As such, geotechnical reports are valuable tools that assist the developer in identifying critical issues on a subject site that may affect the design and construction of a potential development.

Prior to releasing a geotechnical engineering firm to perform subsurface exploration on a potential site, a developer must clearly communicate his/her goals to the geotechnical engineer. This is usually achieved through a scope of work (SOW); however, depending upon available information, the developer may be required to rely on the geotechnical engineer's prior experience to develop a suitable SOW. Additionally, the stage in the development process directly impacts the SOW and nature of the geotechnical report. For example, if the developer is engaged in due diligence activities and trying to identify a suitable project site, but has not yet finalized design, then a Preliminary Geotechnical Report (otherwise known as a geotechnical feasibility study) is likely to be the most appropriate choice. Alternatively, if the developer already owns the land and has full project approvals, then a full Final Geotechnical Report would likely be more appropriate.

General. As noted above, there are typically two levels of geotechnical reports: preliminary and final. As previously mentioned, the level of detail and sophistication of the report will depend on where a particular project is in the development process. However, certain details should be incorporated into either level of reporting. These include:

- Evaluation of the existing site conditions, prior site use, and the proposed site conditions.

- Detailed scope of work performed by the geotechnical engineer and an explanation of the purpose of the specific exploration.

- Evaluation of the geologic and topographic conditions of the site.

- Description of methods used for exploration, referencing approved American Society for Testing Materials (ASTM) standards (or equivalent).

- Description and results of laboratory testing program performed.

- Boring/testing location plan and stratified boring logs.

Soil borings are the most prevalent method of obtaining subsurface data; however, the geotechnical engineer may recommend performing a litany of other exploration methods including test pitting, the use of hand augers, cone penetration testing (CPT), or other of geophysical exploration techniques. These alternative methods may be more appropriate in the specific geologic conditions anticipated by the geotechnical engineer and may offer an additional level of detail not afforded by standard borings. In some cases, they may also be more economical to perform, which can assist the developer with cost management.

Depending upon the scale of the proposed project, site size, number of borings, and the amount of laboratory testing required, a typical preliminary report can be prepared in three to four weeks, whereas a full geotechnical report will typically require four to six weeks to prepare. Note, however, that these completion time ranges will vary greatly based upon the level of work requested and/or required.

Preliminary Geotechnical Reports. It behooves the developer to provide the highest level of information to the geotechnical engineer prior to performing any exploration. This enables the geotechnical engineer to design the most appropriate subsurface exploration for the proposed development and the ability to perform the most appropriate analysis. However, this may not be possible during the early stages of due diligence if the developer has yet to complete a preliminary design.

As a result, preliminary geotechnical reports are typically performed during the procurement phase of a project as a feasibility study. Typically, a limited number of borings are performed throughout the site to provide a global evaluation of the site, verses exploring specific features. It is not uncommon to perform these explorations long before a site plan has been developed as the information gleaned from this type of exploration can assist the designer in locating structures in the most advantageous or economical areas of the site to minimize costs.

In addition to the general items previously listed, preliminary reports typically also include preliminary recommendations on feasible foundation types, identify construction-related issues related to problematic soils and groundwater, provide insight into the potential issues with rock excavation and blasting, and provide preliminary feasibility of the site for general construction. It is important to note that none of the preliminary report recommendations are meant to serve as final recommendations and ranges of expected values are given, as opposed to absolute values, for issues investigated. This is the unavoidable result of the relative lack of details that are obtained during a preliminary report exploration, which results in a significantly more conservative evaluation of the site.

Final Geotechnical Reports. The Final Geotechnical Report typically cannot be completed until final design of the proposed project is complete or substantially complete. A geotechnical engineer will be hesitant to perform this level of a study until a civil engineer has developed a site plan and the structural engineer has performed the structural design of potential buildings and structures. If these services have not yet reached a meaningful degree of completion, it may still be possible for the geotechnical engineer to continue with the final geotechnical exploration, but doing so will require the engineer to make (and list) a variety of assumptions during the analysis. If the final design varies significantly from the geotechnical engineer's assumptions, the developer will incur additional fees due to the need to revisit recommendations and reanalyze the site. It should also be noted that the more assumptions a geotechnical engineer must make regarding the site, the more conservative the resulting analysis will generally be, which can be costly for the developer.

Ideally, before initiating the final geotechnical exploration, the geotechnical engineer should be provided a detailed understanding of the proposed development in regards to the following:

- Existing and proposed site grading, including any steep slopes or earth retaining structures.

- Structural loading, building usage, and finished floor elevations of proposed structures.

- Roadways and paved parking areas.

- Stormwater management features and other drainage features including ponds, bio-retention facilities, rain gardens, and pervious pavements.

- Underground utilities, culverts, and other ancillary structures.

In addition to the general notes, and those included in the preliminary report, the Final Geotechnical Report should include final design and construction recommendations for the following:

- Building and structure foundations

- Rock excavation

- Groundwater and site dewatering

- Unsuitable soil remediation and/or over-excavation

- Cut/fill slope stability

- Below grade walls and other earth retaining structures

- Stormwater management features and drainage structures

- Dams/embankments

- Pavement and roadways

The report's design and construction recommendations should list options for the design team to consider, where appropriate. This may include options regarding shallow foundations, such as spread footings, or alternatives for deep foundations or ground improvement, if building loads dictate. Recommendations for building foundation options will be narrowed after consultation with the structural engineer regarding building loads and, in some cases, owner preference. Settlement analysis will also be performed, after which final foundation design parameters can be provided.

Importance. Geotechnical reports are important to developers for several reasons. First and foremost, they are required for compliance with the International Building Code (IBC) and are typically required by local jurisdictions as part of the overall site plan approval process. Beyond regulatory mandates, however, a thorough and concise geotechnical report can provide the developer with valuable subsurface information about the subject site, which can help avoid costly design and construction mistakes. Geotechnical studies and subsequent reporting can provide the developer significant cost savings throughout the entire development process and construction.

For example, geotechnical reports can identify areas of unsuitable soil in advance of construction. This allows the developer to evaluate multiple options. A developer may decide to (1) not purchase the site if the implications of remediation are significant, (2) work with the design team to design around unsuitable areas, or (3) accept an increase in project cost to haul away unsuitable soils, or to modify/stabilize unsuitable soils in place. Without a thorough geotechnical study, poor quality soil might not be discovered until after construction begins and can create unexpected costs as well as the potential need to redesign the site. Alternatively, if an area that was originally going to remain undeveloped was left unexplored, but contained significant quantities of quality structural fill material, this area could be exploited by the developer to either (1) sell off to other developers in need of quality soil, or (2) utilize the material on-site for the proposed project as necessary, or reserve it in case of an unanticipated future need for quality structural fill. Performing a few additional soil borings during the geotechnical exploration can help identify these scenarios and potentially provide significant cost savings.

Conversely, if the developer insists on a limited SOW to align with a predetermined budget, the geotechnical engineer may be required to be far more conservative with their foundation recommendations due to lack of reliable subsurface data. This can result in significant additional costs if the resulting report recommends increased spread footing sizes in order to compensate for potentially low strength soils. Worse and more costly, a recommend could be made for deep foundations when a shallow foundation option may have been feasible.

Ultimately, full geotechnical exploration and reporting is an important part of both the due diligence period and the overall development process. Good geotechnical information not only assists with building and site design, but can also significantly reduce potential cost and schedule overruns in both design and construction. The developer may be tempted to save money on the front end of the project by reducing the SOW, but must recognize the importance of geotechnical reporting. Conversely, geotechnical engineers may prefer as comprehensive a SOW as possible without appreciating the developer's financial constraints, especially during the early stages of a proposed project. The developer and his/her geotechnical engineering team members should agree on a reasonable SOW and level of investigation to ensure the best possible outcome. Once completed, geotechnical results should be shared with other members of the development team, especially the design team.

Environmental Review. There are several important environmental studies and activities that must be undertaken during the due diligence period, including a desktop review and constraint mapping exercise, a preliminary site investigation, completion of a Phase I Environmental Site Assessment, and a Wetland Delineation Report. The developer will need one or more specialized environmental consultants in order to complete these tasks. These may be independent experts or divisions of a larger engineering firm. The results of these activities have implications for site layout, project costs, and types of permit applications, as well as potential legal ramifications. Thus, the developer needs to ensure the awareness and involvement (where appropriate) of several different team members including architects, engineers, and attorneys with the work of the environmental consultant.

The environmental consultant begins the study of a subject site with desktop research using mapping software and extensive database reviews to identify and evaluate possible environmental constraints. This information will be used to produce a *constraint map*. The map not only shows locations of environmental interest or concern, but also helps identify what types of permits will be needed. In most cases, portions of the constraint map can also be used in some of the developer's permit applications. Specifically, the constraint map includes the following:

- Aerial imaging as well as topographic mapping.

- The existence and location of threatened or endangered species, invasive species, and wildlife and waterfowl refuges.

- The location or potential location of cultural resources, such as archeological sites and architecturally important buildings or structures.

- Bodies of water including ponds, rivers, streams, and other wetlands.

- Agricultural and forest districts, prime farmland, and soil types.

- Floodplains, coastal zones, and coast barriers.

- The location of known hazardous materials.

- The locations of private wells and/or drainage fields.

- Land use.

- Parks and recreation.

- Noise and air quality.

- Information found in a review of previously available site reports, if any.

Constraint maps help facilitate the early detection of environmentally sensitive areas and can be used to guide the design team's preliminary design activities. Indeed, depending on the nature of the findings, environmental constraints can trigger a requirement for a redesign or relocation of project elements if not accounted for early on in the design process. Constraint maps also help determine what unexpected resources may be found on the site. For example, identifying areas with large perennial streams or areas next to tidal water indicate an increased likelihood of finding undocumented cultural (archeological) resources. Finally, constraint maps give the environmental consultant a "preview" of the site and can highlight areas that may require special attention during field visits.

Field visits, which can include identifying and delineating the location of wetlands, are part of the environmental consultant's preliminary site investigation activities. During this review, the consultant will visit the subject site to confirm the accuracy of desktop review findings, investigate any unexpected results highlighted in the constraint map, and look for anything not revealed during the desktop review. A thorough inspection of the site helps the consultant gain a more complete understanding of environmentally sensitive areas and site-specific constraints. In addition to assessing features identified in the constraint map, such as waters, including wetlands and endangered species habitats, attention will also be given to other factors during the preliminary site investigation. The consultant will consider all on-site features such as cultural resources, existing structures, utilities, and evidence of hazardous materials such as underground and/or aboveground storage tanks.

After the site investigation and wetland delineation, two reports must be completed: a Phase I Environmental Site Assessment (Phase I ESA) and a Wetland Delineation Report. The purpose of the Phase I ESA is specifically to identify and evaluate the likelihood of any contamination on the site. These reports are necessary for both greenfield and brownfield sites, but are more likely to produce significant results under brownfield conditions. The Wetland Delineation Report pertains to identifying and preserving wetland environments. While it can apply to brownfield sites, wetland delineation is more likely to be relevant to greenfield developments. Note that if endangered species are discovered on a site, a specialized third report will be completed for submission to the United States Fish and Wildlife Service (USFWS) and use in the permitting process.

Phase I ESA. A Phase I ESA can incorporate work already completed for the constraint map and preliminary site investigation. Note that no samples are taken from the site in order to complete the necessary investigations and certain environmental issues are explicitly excluded from the scope of the Phase I ESA. The report is completed with the aim of identifying recognized environmental conditions (REC) or petroleum products associated with the study area. This includes the

> presence or likely presence of any hazardous substances or petroleum products on a property under conditions that indicate an existing release, a past release, or a material threat of release of any hazardous substances or petroleum products into structures on the property, or into the ground, ground water, or surface water of the property.[2]

Sources of contamination can be either on-site from past or current operations or dumping, but may also come from activity on adjacent sites. In order to fully evaluate the risk of contamination, consultants will place a heightened focus on adjacent properties to address any concern that contaminated materials may have leached onto subject site from another property.

A Phase I ESA report is critical to developers for three reasons. First, the lender will almost certainly require this report as a condition of placing the loan. Second, being aware of contamination prior to acquiring the site gives the developer the option of choosing not to purchase the site, accepting potentially costly remediation obligations, or negotiating with seller with respect to the contamination. The seller of a site found to be contaminated may be willing to perform the necessary remediation or may reduce the purchase price to account for the cost of remediation the developer would have to perform in order to use the site. Third, Phase I ESA reports are important because they ensure that the developer and his/her legal counsel are fully aware of any contamination and can evaluate any implications for inherited or future liability associated with it should the developer choose to acquire the site. Note that, depending on the nature of the results, Phase I ESA report findings can trigger the requirement to perform a second, more detailed study. A Phase II ESA is performed to further investigate the potential presence of a suspected contaminant or to assess whether a release of a contaminant or regulated substance that occurred on the property might have adversely impacted the site soil, groundwater, and/or surface water. For example, a Phase II ESA is necessary any time there is an environmental concern or a suspicion of asbestos or lead-based paint in an existing building or other structure, which has implications for rehab and redevelopment projects. Unlike prior investigations, samples will be taken in order to complete a Phase II ESA. If contamination is found, a Phase III ESA study is used to guide remediation efforts.

Wetlands Delineation Reports. Wetlands are part of the "Waters of the United States" and subject to federal legislation and protections.[3] They fall under the jurisdiction of the U.S. Army Corps of Engineers (USACE), which tracks the locations of "jurisdictional areas" of wetland significance. Specifically, wetlands are

> areas that are periodically or permanently inundated by surface or ground water and support vegetation adapted for life in saturated soil. Wetlands include swamps, marshes, bogs and similar areas.[4]

The purpose of wetland delineation reports is to identify the location of waters of the U.S., including wetlands, on a subject site and communicate that information to the USACE for confirmation. Generating the report requires two phases of investigation: (1) desktop research followed by (2) a physical inspection and survey of the wetlands areas. After being assessed, the boundaries of any wetlands will be physically marked with flags by the environmental professional and then recorded by a surveyor. All pertinent information is then sent to the USACE for verification and inclusion as an official jurisdictional area. The USACE will then send an inspector to confirm the findings. Note that if no wetlands are found or if identified wetlands only require a minor modification to an existing jurisdictional area, a Wetland Memo is used rather than a full delineation report. This information will be used after final design during the permitting process to assign "wetlands credits" to the developer based on the size and length of the existing wetlands.

Wetlands determinations have several consequences for development. Any impacts to confirmed jurisdictional waters, including wetlands, will have to be included in permit applications. The developer will be required to either design the proposed development in such a way as to maintain the existing wetlands areas, or to demonstrate that all efforts have been made and that disturbing the wetlands is unavoidable. Thus, wetlands have obvious implications for preliminary design exercises. Avoiding and/or minimizing impact to wetland areas may be possible with respect to design, but may decrease site usability and increase costs. If the developer must infringe on existing wetlands, then, depending on the size and severity of the infringement, he/she may be required to mitigate the loss to the affected wetlands by paying a wetlands bank to recreate them offsite. This typically requires replacing a larger area than the existing on-site wetlands and can be extremely costly.

5.2.2. Site Design

Design is a fluid element that evolves throughout the development process and becomes more permanent as a proposed project matures. Different components of design can change in response to physical, financial, or even legal pressures. While the design team's ultimate objective is to produce a final design and construction documents, those goals are far from being realized during the due diligence phase. Rather, critical preliminary site design considerations are undertaken during the study period that will eventually allow the developer and team to achieve the final design milestone at a future project stage.

Preliminary Design. *Preliminary design* or *concept design* refers to the first stage of the design team's efforts to translate their developer-client's programmatic thoughts and requirements into a workable solution for a particular site. The main goal of preliminary design is to capture the developer's project vision and confirm that the scope can be achieved. This requires exploring possibilities, new ideas, and concepts in order to achieve the developer's goals. Note that if the developer is a merchant builder working with a build-to-suit client, the client/end-user should also be involved in the design process to make sure key needs are met. Preliminary design exercises apply to the siting of any planned structures or buildings, parking support for the building components, as well as any surrounding support buildings. Another goal of preliminary design is to maximize site resources to serve their highest purpose. This typically applies to maximizing land use but could also include deliberate use of topography, tree and natural vegetation, or other unique site features for the benefit of the project. The end product of a preliminary design exercise is a graphical/visual representation of both the proposed project and any governmental restrictions that must be accommodated on the site in order to implement the project as envisioned. A combination of both technical and creative skills is required to generate a thorough preliminary design and the developer will work closely with his/her architect and civil engineer during this time.

Preliminary design documents and exhibits aid the developer in several ways. First, the design exercise itself helps capture the developer's vision by requiring him/her to articulate goals and describe components of the proposed project. This may include a discussion of client needs if the developer is building for a specific client and, indeed, when appropriate the client may be involved in the dialogue. The descriptive process assists in solidifying project ideas and concepts into something that can be rendered. The resulting design representations allow the developer and the design team to see if and how the proposed project concept may fit on the chosen site. Preliminary designs can also exhibit alternative project elements in order to provide options and allow the developer to explore alternatives. For large sites, preliminary design exercises may also help solidify components of the development's overall master plan.

Next, the preliminary design can be used as the basis for an initial, rough estimate of project costs. Design exhibits and documentation are often given to specialized consultants who can provide preliminary professional cost estimates based on the proposed program. These estimates allow the developer to determine the basic cost ramifications of the project and make informed decisions about the financial viability of the overall project. Note that because the design is not complete and does not contain all project details, the result of the "costing" exercise is also not conclusive. However, cost estimates based on preliminary design are more informed than the developer's "back of the envelope" estimates, especially if the design exercises reveal cost considerations that the developer was not aware of previously. Future sections in this chapter will discuss project financial analysis and the creation of the developer's pro forma. If the estimated project costs are not supportable, the developer and design team may be able to modify the preliminary design in order to make the project viable.

Finally, preliminary design documents and exhibits create a visual representation of the development project that can be used in promotional and marketing materials. These materials may be used to generate interest in the project from potential investors and financial institutions. They can also be used in presentations to communicate the developer's vision and inform governmental agencies and community organizations as needed to pursue approvals and/or zoning changes.

There are several challenges posed by the preliminary design process. Unlike the production of consumer products, whose components are often easily mass-produced and assembled, design products are entirely customized for each project and site. This involves considerable time and effort. Even if two seemingly simple projects are similar, site layout and conditions, zoning restrictions, and other limitations still necessitate a bespoke effort. Further, the sequence of subsequent events between the start of preliminary design exercises and the completion of the project is extensive, making it challenging to successfully design for all future contingencies. Indeed, project designs frequently evolve and change considerably after the preliminary design. This is often the result of concessions made to government and community groups in exchange for project approvals. Note that, while necessary in order to accomplish the project, such changes do not necessarily improve the preliminary design. From the design team's standpoint, design-related activities can be challenging because of their interconnectivity with legal, financial, regulatory, and other project-based activities. This often leads to the need to accommodate a flurry of design activity performed in advance of the developer's deadlines in these other areas. Conversely, there are often periods of very little design development activity while the developer and/or other team members mull over financial or other regulatory issues before deciding to proceed with the next phase of work.

While the architect is usually the most active development team member involved in the creation of a project's preliminary design, input from a range of different disciplines can impact final project results. Thus, best practices suggest that creating a reliable and appropriate preliminary design involves seeking contributions from many different development team members. Discussions between professionals can assist in solidifying needs and restrictive components. Among those that should be included in early consultations are:

- Civil engineers
- Planners
- Landscape architects
- Stormwater management engineers
- Traffic engineers
- City planners
- Municipal zoning and permit group
- Financial and or investment groups
- Marketing specialists

Given the level of effort involved in creating a preliminary design, the developer will most likely put a contract in place for design services either before or during the due diligence period. In the United States, use of the American Institute of Architects (AIA) form contracts is a standard practice for the construction industry; however, as with all contracts, the terms are negotiable and the forms may be modified by the parties involved. Various AIA contracts are available, depending on the nature of the proposed project and scope of design work. Contracts accommodating hourly time and material efforts may be appropriate for use when initial exploratory tasks need to be performed. "Lump sum" contractual documents are used when a clear set of goals are agreed to and confirmed in advance by both the developer-client and the design professional. Other forms of AIA contracts facilitate the phased delivery of portions of work on design documents as needed to accommodate project schedules and milestones. Additionally, large firms or those with long-standing client relationships may have customized contracts for specialty scopes of work.

Infrastructure Considerations and Zoning Constraints.
It is not only the developer's vision for the proposed project that must be included in the preliminary design. Infrastructure considerations and zoning constraints must also be incorporated. Site access is, as always, critical, and the design team will assess the property's access to and from dedicated roadways. Access options may change as

different project elements are included or moved within the design. However, if access points are fixed, the design must be developed around them. This may or may not have consequences for the efficiency of site utilization. Apart from road infrastructure, the availability and adequacy of utilities must also be determined. Utility requirements can affect design if their location or the cost to extend them dictates where buildings are located. In some cases, an especially large or specialized project may require backup power provided by a substation, which must be accommodated on the property. The location of the substation building may have implications for the project and can impact the overall site design. Stormwater control must also be accounted for in preliminary design planning. Modern stormwater guidelines are needed to control water for the surrounding sites, neighborhoods, and downstream entities. However, existing regulations often require large swaths of land to be set aside to accommodate stormwater. If this area cannot be accommodated on a small site, then subsurface retaining structures must be incorporated into the concept design, although this has significant cost implications.

Governmental restrictions, as discussed in Chap. 2, can hamper the use of a parcel of land by limiting the types of allowable uses and/or density needed to make a project viable. During the due diligence period, the developer and the design team must make a final determination with respect to whether or not the current zoning is appropriate for the proposed project. As a matter of practice, the developer should also check if the property benefits from any existing site plan approvals put in place by the current or previous owners. If the developer intends to work within the existing zoning designation, the preliminary design must comply with allowable density, setback requirements, height limitations, and other applicable zoning regulations. Depending on zoning and other governmental restrictions, creative solutions may need to be sought in order to accommodate a project on a site. Certain solutions, such as use of a subsurface parking garage below a building on a small lot, may allow the proposed project to fit on the site, but will lead to a significant increase in project costs. If the preliminary design reveals that the project cannot be accommodated under existing zoning, the developer will either have to reconsider the project vision or else pursue a variance or change of zoning.

The developer may also decide to apply for a change of zoning if he/she wants to pursue a substantially different use than current zoning allows. In any case involving rezoning, the preliminary design must imbed the developer's vision in a plan that can be reasonably supported before a zoning committee. This will depend on the type of zoning regime, surrounding uses, details of the prevailing comprehensive plan, and other factors, all of which the design team will need to consider.

Site Conditions. Site conditions and land elements also impact preliminary design. The results of technical surveys and studies completed during due diligence can be used to inform preliminary design exercises in this regard. However, coordination can be a challenge if technical studies have not yet been completed when the design team is ready to begin work.

Site topography must be mapped and accurate plans provided to the design team in order to determine implications for building placement. Sites with minimal topographic changes can usually be regraded to make a larger portion of the site useable. However, moderate to severe topographic changes may require the use of one or more retaining walls, which can impact project cost and inherently limit site usability. Similarly, environmental or zoning considerations revealed during due diligence, such as protected foliage cover or wetlands, may strictly limit the development of a site. The results of geotechnical due diligence also include pertinent information for design, such as soil reports indicating poor soil that must be removed, at additional cost to the developer, or else avoided in site design. Finally, easements or other considerations found on the title report may have to be accounted for in project design.

Summary/Conclusions. In order to arrive at a viable solution and finished preliminary design, the developer needs the combined expertise of the many development team members, including the architect, civil engineer, surveyors, environmental specialist, geotechnical engineer, zoning consultant, title abstractor, and others. Indeed, the creation of a project's preliminary design is an ideal part of the development process to observe the interdependency of team members and the importance of each discipline to the project as a whole.

5.2.3. Special Considerations of Public Sector Projects

Public sector projects will normally have the same technical due diligence concerns as private sector projects: environmental, topography, soil quality, and the impact on design, timeline, and cost. However, the public sector is often subject to additional reviews, particularly if a state or local entity uses federal funding, such as undertaking highway construction using funding from the Highway Trust Fund. One well-known example of federal due diligence is created by the National Environmental Policy Act (NEPA) of 1969, which specifically applies to the development of large federal projects such as airports, buildings, military complexes, and highways.[5] Projects subject to NEPA review require the developer to produce special environmental assessment reports that include justification of project need, environmental impacts, discussion of alternatives, and sources consulted. The consequences of these reporting requirements on public sector projects will be highlighted further in case studies in Chap. 7.

5.3. MARKET AND FINANCIAL DUE DILIGENCE

The previous chapter described the developer's preliminary market research and basic financial evaluation of potential sites during a site search. While that level of study may have been enough to warrant putting a site under contract as part of achieving Milestone 2 (selecting a site), it does not provide the developer with sufficient information to justify actually purchasing the site. In addition to technical and legal due diligence, the developer will also perform a thorough financial analysis to evaluate the project during the study period as part of Milestone 4 (completing due diligence). This involves creating a pro forma informed by a detailed market feasibility study and cost estimates based on site conditions and preliminary design exercises.

5.3.1. Market Analysis

A *market analysis study*, also sometimes called a *market feasibility study*, is a professional examination of a subject property (or area) in the context of broader market trends. These studies, which are undertaken by specialized real estate analysts, help developers and other real estate professionals understand the fundamental market support (or lack thereof) for different real estate project types on a particular site. Market analysis studies are informed by historical market and demographic data, which is used to make forward-looking projections and recommendations. While final study recommendations are based on quantitative data, they also rely on the professional judgment of the analyst, making the result a classic combination of "art and science." Ultimately, a market analysis study is a form of risk assessment that evaluates alternative potential future outcomes in order to help developers make the best project decisions possible based on available information.

Market analysis studies can be completed by a variety of different firms employing qualified real estate analysts. Large development companies may have the resources to do their market studies in-house while other developers may hire specialized consulting firms to complete their market studies. Some brokerage firms also have analyst divisions that can complete market studies for their clients. Naturally, the time necessary to complete a market study varies based on the project and the land uses being studied. For example, a rental apartment market study can often be completed in two to three weeks, whereas a long-term redevelopment plan may take as long as one year. Typically, however, most market studies can be completed in four to six weeks.

There are several objectives of a typical market study. First, the study seeks to assess the subject site (or area) in terms of location, site characteristics, the proposed site plan, and planned redevelopment potential, among other considerations. It also serves to analyze competitive properties with respect to pricing, size, absorption, market audience, positioning, features, and amenities offered, as well as any other relevant factors. Market studies also assess the growth potential of the market area and the character of employment that will drive demand for future land uses in the area. Another goal of a typical market study is to promote understanding of economic and demographic trends in order to assess the depth of current and future demand in the primary market area. This allows the analyst to determine the likely market demand for the various land uses being envisioned for a particular subject site and, using this information, determine a likely capture and absorption timeline for the subject site. Finally, market analysis studies evaluate the market depth and positioning for the targeted land uses on the subject site. Using the site analysis, competitive market data, the target market audience, and demand forecast, the analyst generates final recommendations, typically including pricing, unit mix, total square feet supported, absorption, and positioning strategy for land uses on the subject site.

Market study reports include five main sections: (1) subject site analysis, (2) economic and demographic analysis, (3) competitive market analysis, (4) demand analysis, and (5) findings and recommendations. The following will provide additional detail on each of these sections.

Subject Site Analysis. The objective of the subject site analysis is to determine the appropriateness of the site for the proposed/potential use (or uses) in the context of possible alternatives. Items examined include characteristics such as property size, topography, layout, and constraints; surrounding land uses; proximity to employment and services; area prestige/reputation; access, visibility, and frontage; and planned infrastructure improvements. The subject site analysis often includes a SWOT analysis:

- *Strengths:* Site/area characteristics that promote development

- *Challenges:* Site/area characteristics that present challenges

- *Opportunities:* Preliminary market opportunities based upon site analysis

- *Threats:* Competitive threat assessment

Economic and Demographic Analysis. The purpose of this section is twofold. The first is to better understand "the big picture" and how macro market elements can impact the subject site. For example, this might include an analysis of information pertaining to larger demographic trends, such as the desires of Millennials versus Baby Boomers, and economic trends, such national-level projections for the economy. The second purpose of this section of the study is to characterize regional socioeconomic trends and conditions, which are currently and will continue to influence short- and long-term development potential.

The analyst provides a discussion of locally and nationally prepared economic forecasts, for the relevant regional and local areas, with particular attention focused on economic growth and subsequent annual growth. This information is critical to a full understanding of the demand potential of the site and some of the information will flow directly into the demand analysis model. Commentary is offered on the reasonableness of available economic geographic projections based on observed local and regional economic patterns; when appropriate, alternative projections are offered along with an explanation of the rationale for these changes. This section also provides a delineation of the primary market area (PMA) and secondary market area (SMA). These are the areas from which most potential buyers/renters/tenants would likely emanate. The PMA varies considerably by product type and is determined by a variety of factors, including access (including drive times) and visibility, location of competitors, target market demographics, etc. The specific demographic characteristics of the target market audience within these areas is also defined. More specifically, the economic, demographic and real estate market information that is analyzed includes, but is certainly not limited to, historical and projected growth patterns of population and households; current distribution of household income, age, and household size; employment trends, income, commuting patterns, and other pertinent factors.

Competitive Market Analysis. The goal of this section is to understand the appropriate positioning and orientation of the subject property compared to competing properties. This section seeks to answer questions such as:

- How is the market segmented?
- What are the dominant consumer preferences?
- What trade-offs exist between location, quality, and product in the marketplace?
- What are the subject site's competitive advantages and disadvantages in this marketplace?

To conduct the competitive market analysis, comparable projects are selected for study from the competitive market area (CMA), which is the geographical area within which the subject site will directly compete for consumers and/or tenants. Or, in other words, the CMA is the area within which the subject site competes on a more or less equal basis with other properties. The CMA area is uniquely defined for each subject site based upon geographical proximity, product type, and market orientation. Note that the CMA is typically smaller than the PMA. This section of the report also typically includes competitive surveys of relevant projects within the CMA and interviews with knowledgeable real estate professionals. Since the market analysis is intended to help the developer evaluate future conditions, which will exist once the proposed project is actually built, identifying other planned and proposed projects in the area is also critical. Recall that this is known as the development "pipeline." Individual case studies may also be included in this section of the report when local comparable projects do not exist. These are very useful if a proposed project is first of its kind in a particular area/region and can help the developer understand market acceptance, critical success factors, critical failures, and lessons learned from the case study experience.

Demand Analysis. This section provides a quantitative, forward-looking context to the competitive market analysis. It is a statistical demand analysis that determines the potential of various land uses by market sector and customer type, a projection of supportable new development in the relevant trade/catchment area, and the best

opportunity and positioning for the future developments relative to the market and key competitors. This section also utilizes *capture rates*, which refer to a proposed project's expected capture of future demand. This is determined based on historical capture trends, projected supply, competitive advantages and challenges of the site, and uniqueness of the project. One metric commonly used to determine capture rate is known as "fair share," which is the project's fair share capture of future demand. For example, if there are five projects in the PMA including the subject property, the subject property's "fair share" capture of demand would be 20 percent. Often, however, new projects garner more than their fair share of demand if they have unique features, benefit from a superior location, or because tenants may prefer a new building. Therefore, factors other than "fair share" must be considered when calculating a capture rate.

This demand information can help the developer identify underserved markets where unmet demand may exist, which could represent an opportunity for the developer. However, the findings of this section are only as useful as the quality of the underlying demographic data and information on consumer preferences used in the analysis. For this reason, it is important for the analyst to consult current and reliable sources whenever possible and make note of any limiting information. Further, results will be dependent on the assumptions used in the analysis, which should be clearly defined.

Findings and Recommendations. This section ties all the pieces of the market study together and connects the analysis to a recommendation(s). Recommendations are data-driven and explained by a clear analytical commentary. Specific recommendations, each of which is vitally important for the developer and the development team to consider, are made on the following types of information:

- Overall suitability of the site for the proposed land uses (or product types)
- Highest and best land use(s)
- Positioning for prices/rents

- Absorption/sales
- Total supportable units/square feet
- Orientation and target consumer
- Supportable features and amenities

The information contained in most studies is valuable for several different members of the real estate industry including developers, builders, landowners, investors, banks and lenders, architects, and planners and municipalities. For example, municipalities might use study information to inform long-range planning exercises, particularly with respect to stimulating revitalization in target areas. While market analysis studies are useful to many different real estate professionals, they are especially important for developers. Although a market study can be done on a general area to support a site search, developers who do not have this capability internally will usually only contract to have a market study completed after they have identified and gained control of a specific site. As with other studies discussed in this chapter, developers may be unwilling to commit financial resources toward a market study until the site is under contract and the due diligence period creates both the time and need for the study to be completed. During this time, the developer needs to better understand the project's potential and, depending on the degree of viability, whether or not it makes sense to purchase the property. If the developer does not already have a specific project concept in mind for the site, the market study will help determine the highest and best use of the site and identify which product types are appropriate for the market. If the developer has a proposed project in mind, the study will help verify whether or not the proposed product/program is supportable. In all cases, the study will provide the developer with achievable rental or sales pricing and absorption rates.

The developer and the development team can use the results of a market analysis study in several ways. The demand, pricing, and absorption information is often utilized in a financial pro forma to help the developer understand if the project makes sense financially. The design/engineering team can make design recommendations based on the market trends, comparable properties, and demand information. The lender may use the study in the underwriting process to help evaluate the risk of a proposed project before providing financing.

Developers may also rely on a market analysis report when contemplating a redevelopment or repositioning project in which an existing building has potentially outlived its useful life or is under-performing, such as may be the case with an aging shopping center or an outdated apartment complex. In these instances, a market study can help determine if the market exists for an updated and costlier product. Finally, a market study might be utilized to help determine the value of the property before disposition, or sale.

5.3.2. Developing the Pro Forma

Prior to the due diligence period, the developer did not have the site information, design vision, and completed market study necessary to make a detailed evaluation of the proposed project. Before deciding to purchase the site, the developer will use the due diligence information gathered from various team members to create a detailed financial model known as a pro forma. A *pro forma* is a customizable financial tool that allows the developer to conduct a thorough evaluation of project revenue, expenses, and returns to determine if a proposed project will be profitable and if the amount of profit is sufficient to compensate the developer and any equity investors. At its most basic level, a pro forma reconciles all of the anticipated annual revenue with the expected annual costs of a proposed project to indicate the amount of profit or loss the project will generate each year. However, this same information is used for several other purposes, including determining: (1) whether or not project revenues are sufficient to cover debt service, (2) if amenities, finishes, and architectural aspirations are supported at the proposed level or require value engineering, (3) if site layout must be changed in order to accommodate additional leasable space, (4) if the project will satisfy the required rate of return for the developer and equity investors, and (5) the total amount of the investment the developer should be willing to make in order undertake the project.

The following sections will discuss the basic pro forma assumptions, inputs, and measures of returns in the context of a hypothetical mixed-use development consisting of multifamily apartments and first-floor retail space; for this scenario, we assume the Atherton Road site is a 6-acre site with appropriate zoning in place. The level of detail provided is not intended to allow development team members to undertake a financial analysis, but rather to gain a basic understanding of the process and issues involved. This may help architects and civil engineers understand why certain design, site layout, construction, or other decisions are made by their developer-clients.

Assumptions and Inputs. The value of a pro forma analysis is only as reliable as the quality of the information used. The most important considerations for the developer when setting up a pro forma is making realistic assumptions and using accurate, reliable input data. If the developer uses inaccurate data or inflates assumptions, the pro forma results will be misleading. *Inputs* refer to information that must be included in the pro forma in order to analyze the proposed project; the total size of the project in square feet is an example of an input. However, while some inputs are known with relative certainty, many are not knowable in advance. *Assumptions* refer to the developer's decisions about what input data is reasonable to use. For example, the inflationary increase in total utility costs five years into the future cannot be known for certain, often even if utility contracts are in-place, so the developer will make an assumption about future inflation rates for use in the pro forma. This section will describe some of the main assumptions and inputs used in development pro formas.

Investment Horizon. An important assumption that guides the nature of the developer's relationship with the project is the investment horizon. An *investment horizon*, or *holding period*, is the length of time the developer anticipates retaining the asset after construction; it is used in the pro forma as the analysis period or, in other words, the number of years of operations over which financial projections will be run. Investment horizon assumptions are based on the developer's preferences and

business model. Merchant builders will have a relatively short investment horizon, perhaps less than a single year, compared to a legacy developer that intends to retain ownership of the property for many decades. However, few developers can predict the actual length of investment with certainty. A merchant builder that is unable to find a buyer for a newly completed building may be forced to retain ownership of the asset for several years. A developer that intends to hold an asset for many years may be forced to sell due to unexpected circumstances or may make a deliberate decision to sell in order to capture favorable market conditions and achieve a higher-than-expected sales price. Regardless of the type of developer, 10 years of postdevelopment operations is an industry standard period of pro forma analysis and is often used to demonstrate a project's financial viability in loan packages or when seeking equity investors. Note that pro formas always extend for one additional year beyond the investment horizon, meaning that a 10-year pro forma will actually show 11 years of cash flows. This is because the developer, investors, and lenders need to know what the anticipated cash flow will be in the year of sale (the 11th year) in order to calculate the sale value, or *residual value*, of the project for that year.

Construction Costs. There are a range of project costs that are specific to the development of the site itself that must be included in the pro forma. Recall from Chap. 3 that these are generally broken down into three categories: hard costs, soft costs, and land costs. Hard costs are tangible items directly related to the construction of the building as well as any associated labor costs of construction and contingencies for hard costs overruns. Soft costs are necessary but indirect costs required to complete the project. Whereas the developer previously used rough estimates to evaluate these costs as a single line item, he/she will now endeavor to assign more accurate values to each individual category for use in the pro forma. For example, the developer may now be able to include specific quotes for attorney services and architectural and engineering contracts. Indeed, each of these team members may be already be working with the developer in an official capacity during the due diligence period.

As the developer receives site and design information from due diligence exercises, he/she will also be able to refine hard cost inputs. For example, after reviewing the market study and establishing a preliminary design, the developer will know what amenities should be offered, the quality and type of materials or finishes necessary, and the estimated total size of space in square feet that can be accommodated on the site. These details impact the cost of construction and length of time necessary to complete the project. However, it is important to note that material and construction costs based on the project's preliminary design have not yet been validated and may change as the project design is further refined; at this stage, such cost estimates are known as *feasibility cost estimates*. Further, many early construction cost inputs are based on the assumption that the developer can obtain the necessary approvals in a timely fashion. If there is a long delay in the approval process, material costs may increase between the time the developer creates the pro forma and the time construction actually begins. To account for this possibility, the developer may decide to include a premium in the construction input costs or add an overall contingency. Figure 5.1 provides an example of pro forma cost inputs.

In addition to determining what construction costs will be, the developer must also assess when the costs will be incurred. Payments on construction contracts and for development team member services are not made in bulk, but rather as work is completed. The timing of these payments is important because, to a large degree, they will dictate when the developer needs to access funds, or *draw*, from a construction loan. To evaluate the timing implications, the pro forma includes a *construction schedule*, also sometimes called a *development budget*, that translates total cost data into anticipated monthly development expenses. Figure 5.2 shows an example of the first six months of a construction cost schedule. Note that a full pro forma will include monthly costs for the entire construction period, usually estimated to be 24 months for typical projects.

Loan Information.   Typically, no loans have been placed during the due diligence period. This means the developer must make assumptions about loan inputs such as the loan to value ratio (LTV) and the interest rate used to calculate the debt service in the pro forma. However, the developer will have begun discussions with lenders as part of his/her due diligence and should have reasonable confidence in what loan terms will be offered. Once a loan has been issued, the developer can use the actual bank terms, such as interest rate (if it is different than the input used), to adjust the pro forma calculations. Figure 5.3 provides an example of typical loan inputs used in a development pro forma. More details about debt financing will be explained in subsequent sections of this chapter.

Revenue and Expenses from Operations.   As with construction costs, the developer will break out individual categories in the pro forma for both income line items as well as operating expenses. For example, the developer will now have a more accurate idea of total rentable square footage based on site conditions and preliminary design as well as more accurate rental figures based on the market study, both of which inform potential gross income (PGI). In the case of a multifamily project, the developer will decide on the *unit mix*, which is the number and size of different apartment units that will be offered. Additionally, in mixed-use development pro formas, separate revenue line items will be expressed for each different product type represented on the site. In the current mixed-use Atherton Road example, the pro forma includes residential rental income from the lease of multifamily apartments and also commercial rental income from the lease of first floor retail space, as shown in Fig. 5.4. Concessions, free rent periods, vacancies, and debt collection losses will also be individually reflected in the pro forma order to determine effective gross income (EGI). The developer will also include detail for other sources of revenue when calculating EGI, such as fees from renting out parking spaces or storage units, vending income from third-party vendors offering on-site conveniences, or income from leasing rooftop space to wireless service providers to place antennas or other telecommunications hardware. Overall, the revenue inputs in the pro forma should be more detailed

COST INPUTS
Inputs are in BLUE

Project Cost Assumptions

Land Purchase

Purchase Price	$17,235,000
State Recordation Costs	$43,088
County Recordation Costs	$14,357
Grantors Tax	$17,235
Regional Congestion Relief Fee	$25,853
-	$0
-	$0
-	$0
Total Land Purchase	**$17,335,532**

Construction Costs

Residential Construction Cost	$64,350,700
Commercial Construction Costs	$1,800,000
Utilities	$250,000
Landscaping	$150,000
Irrigation	$25,000
Signage	$100,000
Entrance Sign	$25,000
Appliances	$1,340,500
Total Construction Costs	**$68,041,200**

Architectural and Engineering

Architectural	$2,872,500
Engineering	$200,000
Wetland Consultant	$25,000
Traffic Engineer	$35,000
Utility Consultant	$75,000
Signage Design	$100,000
Landscape Architect (tree removal)	$100,000
Geotechnical	$25,000
Testing and Inspections	$250,000
-	
Other	$250,000
Total Architectural and Engineering	**$3,932,500**

Permits andFees

Building Permits	$50,000
Site Plan Permits	$50,000
Bonds	$25,000
Utility Fees	$250,000
Water Tap Fees	$885,600
Sewer Tap Fees	$2,092,500
Proferrs	$4,787,500
Total Permits and Fees	**$8,140,600**

Construction Administration (As a Percentage of Construction Cost)

Development Fee, %	4%
Development Fee	$4,585,543
Construction Management Fee, %	1%
Construction Management Fee	$680,412
Total Construction Administration	$5,265,955

Finance, Legal and Insurance

State Recordation Costs	$186,288
County Recordation Costs	$62,071
Bank Fee	$372,575
Legal Lender	$1,00,000
Legal Loan	$1,00,000
Legal Land	$25,000
Legal Contracts	$1,00,000
Legal Partnership	$75,000
Builders Risk Insurance	$2,50,000
Liability Insurance	$75,000
Title Insurance Other	$1,75,000
-	$0
Total Finance and Legal	**$1,520,934**

Real Estate Taxes

Total Project Cost	$114,638,584
Total Tax Rate Per $100	1.090
Total Ad Valorem Taxes	$1,249,561
Divided by 2	$624,780
Divided by 12 Months	$52,065
Construction Duration in Months	24
Total Real Estate Taxes	$1,249,561

Contingency

Hard Cost Contingency	5.00%	$3,402,060
Soft Cost Contingency	5.00%	$846,245
Total Contingency (% of TDC)	3.71%	$4,248,305

Marketing, Advertising and Commissions

Marketing	$4,21,300
Advertising	$50,000
Commissions, %	4.50%
Lease Rate	$45.00
Term, Years	7
Annual Rent Escalation	1.00%
Leasing Commissions	$260,000
TI Allowance PSF	$75.00
TI Allowance	$1,350,000
Other	$0.00
Total Marketing, Advertising and Commissions	**$2,081,300**

Construction Loan Interest Expense — Debt

LTC		65.00%
Loan Amount		$74,515,080
Term	Years	15
I/O	Years	2
Construction Interest Expense		4.50%
Construction Period Interest		$2,822,697

Total Development Cost	**$114,638,584**

FIGURE 5.1 Pro forma cost inputs.

DEVELOPMENT BUDGET

Inputs are in BLUE

		%	CLOSING 0.00%	1.00%	1.50%	2.00%	2.00%	2.00%	2.50%
Construction Cost Curve		%	0.00%	1.00%	1.50%	2.00%	2.00%	2.00%	2.50%
Year		Paid	-	Sep-18	Oct-18	Nov-18	Dec-18	Jan-19	Feb-19
Period		at	0	1	2	3	4	5	6
Construction Period - Months		Closing	24						
Land Purchase	$17,335,532	100%	$17,335,532	$-	$-	$-	$-	$-	$-
Construction Costs	$68,041,200		$-	$680,412	$1,020,618	$1,360,824	$1,360,824	$1,360,824	$1,701,030
Architectural and Engineering	$3,932,500	50%	$1,966,250	$81,927	$81,927	$81,927	$81,927	$81,927	$81,927
Permits and Fees	$8,140,600	50%	$4,070,300	$169,596	$169,596	$169,596	$169,596	$169,596	$169,596
Finance, Legal and Insurance	$1,520,934	75%	$1,140,701	$15,843	$15,843	$15,843	$15,843	$15,843	$15,843
Construction Administration	$5,265,955	25%	$1,316,489	$164,561	$164,561	$164,561	$164,561	$164,561	$164,561
Real Estate Taxes	$1,249,561		$-	$52,065	$52,065	$52,065	$52,065	$52,065	$52,065
Marketing, Advertising, Commissions	$2,081,300	30%	$624,390	$60,705	$60,705	$60,705	$60,705	$60,705	$60,705
Contingency	$4,248,305	0%	$-	$177,013	$177,013	$177,013	$177,013	$177,013	$177,013
Interest Carry	$2,822,697		$-	$-	$-	$-	$-	$-	$-
Total Land Development	$114,638,584		$26,453,661	$1,402,121	$1,742,327	$2,082,533	$2,082,533	$2,082,533	$2,422,739
Interest Expense				$27,855,783	$29,598,110	$31,680,643	$33,763,177	$35,845,710	$38,268,450

FIGURE 5.2 Sample construction cost schedule.

and more accurate than the estimates used during the "back of the envelope" analysis. Note that the pro forma example in Fig. 5.4 includes "ADUs," or *affordable dwelling units*, in the unit mix. These are income-restricted units that are required by zoning ordinances in many jurisdictions as part of the public sector's effort to address housing affordability. Unlike market rate units, the developer's maximum achievable rent per affordable dwelling unit is capped by a certain percentage of the *area median income* (AMI).

As with income inputs, the developer will include a detailed breakdown of operating expenses. Recall from Chap. 3 that "OpEx" costs were represented by a single, all-inclusive line item for purposes of the preliminary analysis. For the pro forma, however, the developer will create a separate line item for each category, such as costs of advertising

LOAN INPUTS

Inputs are in BLUE

Permanent Debt Service Assumptions

		Percentage of total
Loan Amount	$74,515,080	65.00%
Owner Equity	$4,012,351	3.50%
Equity Partner 1	$18,055,577	15.75%
Equity Partner 2	$18,055,577	15.75%
Total Equity	**$40,123,504**	**35.00%**
Total Funding Sources	**$114,638,584**	**100.00%**
Total Project Costs	**$114,638,584**	

Loan to Value	65%
Interest Rate	4.50%
Loan Term, Years	15
Interest Only Period, Years	2
Amortization Period, Years	30
Bank Fee	0.50%

FIGURE 5.3 Pro forma loan inputs.

INCOME INPUTS
Inputs are in BLUE

Residential Unit Growth Variable and Vacancy Rates

Annual Income Growth - Residential	3%		Stabilized Vacancy % - Residential	5%
			Bad Debts	1%

Residential Unit Income

Unit Mix	Number of Units	NRSF/Unit	Total NRSF	Rent Per NRSF	Rent Per Month	Annual Rent
Studio	15	697	10,455	$2.57	$26,869	$322,432
1BR	175	730	127,750	$2.74	$350,035	$4,200,420
2BR	105	1,135	119,175	$2.29	$272,911	$3,274,929
ADU 60% AMI	11	927	10,197	$1.50	$15,296	$183,546
ADU 70% AMI	11	927	10,197	$1.75	$17,845	$214,137
ADU 80% AMI	22	927	20,394	$2.00	$40,788	$489,456
ADU 100% AMI	22	927	20,394	$2.45	$49,965	$599,584
ADU 120% AMI	22	927	20,394	$2.45	$49,965	$599,584
	-	-	-		$-	$-
Average/Total	383	885	338,956	$2.43	$823,674	$9,884,087
				Average Rent Per Month	$2,151	

Commercial Rental Income

Retail Space	Lease Term	SF/Unit	Rent Per SF (NNN)	Rent Per Month	Annual Rent
Space 1	7	1,800	$45.00	$6,750	$81,000
Space 2	7	1,800	$45.00	$6,750	$81,000
Space 3	7	1,800	$45.00	$6,750	$81,000
Space 4	7	1,800	$45.00	$6,750	$81,000
Space 5	7	1,800	$45.00	$6,750	$81,000
Space 6	7	1,800	$45.00	$6,750	$81,000
Space 7	7	1,800	$45.00	$6,750	$81,000
Space 8	7	1,800	$45.00	$6,750	$81,000
Space 9	7	1,800	$45.00	$6,750	$81,000
Space 10	7	1,800	$45.00	$6,750	$81,000
Average/Total	-	18,000	45.00	67,500	810,000

FIGURE 5.4   Pro forma income assumption inputs.

the property for lease, repair and maintenance costs, landscaping costs, utilities, insurance premiums, staff salaries, taxes, and management fees. The data used for each of these inputs will be as accurate as possible so the developer can make a realistic determination about the project's viability. Figure 5.5 shows pro forma expense assumption inputs.

Inflation.   The inflation rate for costs as well as increases in rental income is an important factor for all pro formas. The U.S. Bureau of Labor Statistics provides 10-year historical data for the consumer price index and other measures of inflation, which can be used to inform pro forma assumptions. More sophisticated pro forma models are able to bound inflation assumptions within a particular range and run several iterations of an investment scenario to reflect a variety of possible future outcomes, including a statistically most likely outcome; these models are said to be *dynamic* because they are capable of reflecting a complex and varying set of potential outcomes. Note that dynamic models are not limited to inflationary projections and can run simulations that manipulate multiple assumptions and inputs simultaneously. However,

EXPENSE INPUTS

Inputs are in BLUE

Expense Variable

Annual Expense Growth	3%

Operating Expense

Administration Expense	$271	Per Unit/Yr	$103,793	Total Yr
Advertising and Promotion	$256	Per Unit/Yr	$98,048	Total Yr
Repairs & Maintenance	$300	Per Unit/Yr	$114,900	Total Yr
Personnel	$1,878	Per Unit/Yr	$719,274	Total Yr
Utilities	$528	Per Unit/Yr	$202,224	Total Yr
- Electric (Common Area Only)	$17	Per Unit/Mo	$78,132	Total Yr
- Gas (Common Area Only)	$8	Per Unit/Mo	$36,768	Total Yr
- Water and Sewer	$11	Per Unit/Mo	$50,556	Total Yr
- Water and Sewer Sub Meter Recapture	0%	%	$-	Total Yr
- Trash Removal	$8	Per Unit/Mo	$36,768	Total Yr

Ad Valorem Taxes

- City Tax Rate		1.09	Per $100 Assessed Value
- County Tax Rate		-	Per $100 Assessed Value
- Other Tax Rate		-	Per $100 Assessed Value

- Total Tax Rate	NOI YR 1	Blended Cap Rate	1.09	Per $100 Assessed Value
Valuation based on NOI and Cap Rate	$7,941,303	5.57%	$142,650,262	
Assessment %			90%	Assessment %
Subtotal Assess Value			$128,385,236	
Total Tax Rate / $100			0.0109	
Total Ad Valorem Taxes			$1,399,399	Total Yr

Insurance	$25	Per Unit/Mo	$114,900	Total Yr
Management Fee (% of Effective Revenues)	$1,083	Per Unit/Yr	$414,691	Total Yr
Common Area Maintenance Recovery (CAM)	$(6)	Per Commercial/SF	$(108,000)	Total Yr
Ground Lease Payment	$-	Per Year	2%	% Escalation Per YR
Replacement Reserves	$250	Per Unit/Yr	$95,750	Total Yr

Operating Expense Summary

Description	Total	Per Unit	Per NRSF
Administration Expense	$103,793	$271	$0.31
Advertising and Promotion	$98,048	$256	$0.29
Repairs & Maintenance	$114,900	$300	$0.34
Personnel	$719,274	$1,878	$2.12
Utilities	$202,224	$528	$0.60
Ad Valorem Taxes	$1,399,399	$3,654	$4.13
Insurance	$114,900	$300	$0.34
Management Fee	$414,691	$1,083	$1.22
Common Area Maintenance (CAM) Recovery	$(108,000)	$(282)	$(0.32)
Ground Lease Payment	$-	$-	$-
Replacement Reserves	$95,750	$250	$0.28
Operating Expenses	**$3,154,979**	$8,238	$9.31

FIGURE 5.5 Pro forma expense assumption inputs.

many developers simply use a 10-year average or will default to a somewhat arbitrary standard assumption to represent a "worst case" inflationary scenario, such as 3 percent. Pro formas that reflect only a single possible outcome are known as *static* financial models. To account for real world variability, these models will sometimes include a *sensitivity analysis* matrix showing the effects of specific changes to inputs or assumptions. Whether there is an appreciable advantage to using dynamic versus static financial models depends on the size and complexity of the project, acceptable margins of error, investor sophistication, and developer preferences.

Outputs. Once all the input data is entered, the pro forma calculates the results. Development pro formas create two important output schedules: the construction draw schedule and the operating schedule. The construction draw schedule shows the timing and amount of draws against the construction loan and the amount of accumulated interest. The developer will not make any loan payments during the construction period, so the interest on loaned funds accumulates during this time. Figure 5.6 shows the first year of a construction draw schedule (note that the construction period often extends beyond a single year). At the end of the construction period, the remaining loan funds available will be $0 and the developer will be obligated to repay the entire construction loan amount plus all accumulated interest. This is either done by replacing the construction loan with a new permanent loan or, in some cases, by converting the construction loan into a permanent loan.

Once construction is complete, the newly created space is occupied and the asset begins operations. To model this in the pro forma, the operating schedule shows the annual cash flow from operations for the period of the investment horizon plus the year of the anticipated sale. Although the developer's entire financial model is referred to as a pro forma, the name "pro forma" or "10-year pro forma" is often also given to the operating schedule itself, perhaps because it effectively summarizes the entire project in a single place. The operating schedule, or "pro forma", will give the developer his/her first detailed assessment of project viability, as shown in Fig. 5.7. If important cash flow measures such as the net operating income (NOI) or before tax cash flow (BTCF) are negative, the project is clearly not viable as proposed. If the projected cash flow loss is substantial, the project will be abandoned and the developer will return to searching for sites; money spend on due diligence activities, studies, and reports, will not be recuperated.

However, it is sometimes possible to revise a nonperforming project concept to make it viable. For example, the developer may explore value engineering to reduce project costs or try to increase rentable square footage. From the development team's perspective, such requests to change project scope or design may appear arbitrary, yet they may be necessary in order for the project to move forward. When reworking a project concept, the developer must avoid the temptation to introduce unrealistic assumptions in order to generate a positive pro forma cash flow. Both artificially increasing asking rents beyond what the market will support and underestimating operating expenses will improve the cash flow shown in the pro forma, but will not actually be representative of achievable project outcomes. In addition to creating false expectations for the developer, circulating an inflated pro forma to potential equity investors and/or lenders can damage the developer's reputation as a reliable professional.

Even when various cash flow measures reflected in the pro forma are positive, the developer will still need to perform additional analysis. In order for a project to be viable, cash flows must not only be positive but they must also be positive by a large enough margin to support the required measures of return for both the developer and equity investors. The next section will discuss how pro forma data is analyzed to evaluate investment returns.

Discounting Cash Flow and Measures of Return.
As a best practice, developers usually look at more than one measure of return to evaluate a project's viability. There are many ways to evaluate the profitability of a development project. For purposes of this book, the most important of these are net present value (NPV), the internal rate of return

DRAW SCHEDULE

Inputs are in BLUE

	CLOSING	Sep-18	Oct-18	Nov-18	Dec-18	Jan-19	Feb-19	Mar-19	Apr-19	May-19	Jun-19	Jul-19	Aug-19
Year	-												
Period	0	1	2	3	4	5	6	7	8	9	10	11	12

Loan Amount

	CLOSING	Sep-18	Oct-18	Nov-18	Dec-18	Jan-19	Feb-19	Mar-19	Apr-19	May-19	Jun-19	Jul-19	Aug-19
Loan Funds Available For Draw	*$74,515,080*	*$74,515,080*	*$74,515,080*	*$74,515,080*	*$74,515,080*	*$74,515,080*	*$74,515,080*	*$74,515,080*	*$72,246,365*	*$68,114,088*	*$62,265,285*	*$54,693,519*	*$45,392,329*
Beginning Loan Draw Balance		$-	$-	$-	$-	$-	$-	$-	$2,268,715	$6,400,992	$12,249,795	$19,821,561	$29,122,751
Construction Loan Draw (65.0%)		$-	$-	$-	$-	$-	$-	$2,268,715	$4,132,277	$5,848,803	$7,571,766	$9,301,190	$7,635,040
Interest Expense (Rate 4.5%)		$-	$-	$-	$-	$-	$-	$-	$8,508	$24,004	$45,937	$74,331	$109,210
Cumulative Draw Balance		$-	$-	$-	$-	$-	$-	$2,268,715	$6,400,992	$12,249,795	$19,821,561	$29,122,751	$36,757,791
Remaining Loan Funds Available	*$74,515,080*	*$74,515,080*	*$74,515,080*	*$74,515,080*	*$74,515,080*	*$74,515,080*	*$74,515,080*	*$72,246,365*	*$68,114,088*	*$62,265,285*	*$54,693,519*	*$45,392,329*	*$37,757,289*

FIGURE 5.6 Sample construction draw schedule.

TEN YEAR PRO FORMA											SALE
Year	1	2	3	4	5	6	7	8	9	10	11
Gross Revenues:											
Residential Rental Revenues	$9,884,087	$10,180,610	$10,486,028	$10,800,609	$11,124,627	$11,458,366	$11,802,117	$12,156,181	$12,520,866	$12,896,492	$13,283,387
Gross Rent per SF / Mo	*$2.43*	*$2.50*	*$2.58*	*$2.66*	*$2.74*	*$2.82*	*$2.90*	*$2.99*	*$3.08*	*$3.17*	*$3.27*
Garage/Carport/Storage Income	$473,705	$487,916	$502,553	$517,630	$533,159	$549,153	$565,628	$582,597	$600,075	$618,077	$636,619
Other Income	$602,536	$620,612	$639,230	$658,407	$678,159	$698,504	$719,459	$741,043	$763,274	$786,172	$809,757
Bad Debts Expense	$(98,841)	$(101,806)	$(104,860)	$(108,006)	$(111,246)	$(114,584)	$(118,021)	$(121,562)	$(125,209)	$(128,965)	$(132,834)
Vacancy	$(494,204)	$(509,031)	$(524,301)	$(540,030)	$(556,231)	$(572,918)	$(590,106)	$(607,809)	$(626,043)	$(644,825)	$(664,169)
Effective Residential Income	$10,367,282	$10,678,301	$10,998,650	$11,328,609	$11,668,468	$12,018,522	$12,379,077	$12,750,450	$13,132,963	$13,526,952	$13,932,760
Effective Rent per SF / Mo	*$2.55*	*$2.63*	*$2.70*	*$2.79*	*$2.87*	*$2.95*	*$3.04*	*$3.13*	*$3.23*	*$3.33*	*$3.43*
Commercial Retail Income (NNN)	$810,000	$818,100	$826,281	$834,544	$842,889	$851,318	$859,831	$868,430	$877,114	$885,885	$894,744
Gross Rent per SF / Yr	*$45.00*	*$45.45*	*$45.90*	*$46.36*	*$46.83*	*$47.30*	*$47.77*	*$48.25*	*$48.73*	*$49.22*	*$49.71*
Vacancy	$(81,000)	$(81,810)	$(82,628)	$(83,454)	$(84,289)	$(85,132)	$(85,983)	$(86,843)	$(87,711)	$(88,589)	$(89,474)
Effct. Commercial Retail Income	$729,000	$736,290	$743,653	$751,089	$758,600	$766,186	$773,848	$781,587	$789,403	$797,297	$805,270
Total Rental Income (EGI)	$11,096,282	$11,414,591	$11,742,303	$12,079,699	$12,427,068	$12,784,708	$13,152,925	$13,532,036	$13,922,366	$14,324,248	$14,738,030
Expenses:											
Administration Expense	$103,793	$106,907	$110,114	$113,417	$116,820	$120,325	$123,934	$127,652	$131,482	$135,426	$139,489
Advertising and Promotion	$98,048	$100,989	$104,019	$107,140	$110,354	$113,665	$117,074	$120,587	$124,204	$127,930	$131,768
Repairs & Maintenance	$114,900	$118,347	$121,897	$125,554	$129,321	$133,201	$137,197	$141,313	$145,552	$149,918	$154,416
Personnel	$719,274	$740,852	$763,078	$785,970	$809,549	$833,836	$858,851	$884,616	$911,155	$938,489	$966,644
Utilities	$202,224	$208,291	$214,539	$220,976	$227,605	$234,433	$241,466	$248,710	$256,171	$263,856	$271,772
Ad Valorem Taxes	$1,399,399	$1,441,381	$1,484,622	$1,529,161	$1,575,036	$1,622,287	$1,670,956	$1,721,084	$1,772,717	$1,825,898	$1,880,675
Insurance	$114,900	$118,347	$121,897	$125,554	$129,321	$133,201	$137,197	$141,313	$145,552	$149,918	$154,416
Management Fee	$414,691	$456,584	$469,692	$483,188	$497,083	$511,388	$526,117	$541,281	$556,895	$572,970	$589,521
CAM Recovery	$(108,000)	$(111,240)	$(114,577)	$(118,015)	$(121,555)	$(125,202)	$(128,958)	$(132,826)	$(136,811)	$(140,916)	$(145,143)
Ground Lease Payment	$-	$-	$-	$-	$-	$-	$-	$-	$-	$-	$-
Replacement Reserves	$95,750	$98,623	$101,581	$104,629	$107,767	$111,000	$114,331	$117,760	$121,293	$124,932	$128,680
Operating Expenses	$3,154,979	$3,279,080	$3,376,864	$3,477,575	$3,581,301	$3,688,133	$3,798,164	$3,911,490	$4,028,210	$4,148,424	$4,272,239
Operating Cost per Unit	*$8,238*	*$8,562*	*$8,817*	*$9,080*	*$9,351*	*$9,630*	*$9,917*	*$10,213*	*$10,518*	*$10,831*	*$11,155*
Net Operating Income	$7,941,303	$8,135,510	$8,365,439	$8,602,124	$8,845,767	$9,096,575	$9,354,761	$9,620,546	$9,894,156	$10,175,824	$10,465,791
Debt Service:											
Interest Expense	$(3,353,179)	$(3,298,215)	$(3,240,778)	$(3,180,756)	$(3,118,033)	$(3,052,488)	$(2,983,993)	$(2,912,416)	$(2,837,618)	$(2,759,454)	$(2,677,772)
Principal Payment	$(1,221,417)	$(1,276,381)	$(1,333,818)	$(1,393,840)	$(1,456,563)	$(1,522,108)	$(1,590,603)	$(1,662,180)	$(1,736,978)	$(1,815,142)	$(1,896,823)
Total Debt Service	$(4,574,596)	$(4,574,596)	$(4,574,596)	$(4,574,596)	$(4,574,596)	$(4,574,596)	$(4,574,596)	$(4,574,596)	$(4,574,596)	$(4,574,596)	$(4,574,596)
DSCR	1.74	1.78	1.83	1.88	1.93	1.99	2.04	2.10	2.16	2.22	2.29
DSCR (on Residential NOI Only)	1.55	1.59	1.64	1.69	1.74	1.79	1.85	1.90	1.96	2.02	2.08
Net Cash Flow From Operations	$3,366,707	$3,560,915	$3,790,843	$4,027,528	$4,271,171	$4,521,979	$4,780,165	$5,045,950	$5,319,560	$5,601,228	$5,891,195
Cash Flow Summary											
Net Operating Income	$7,941,303	$8,135,510	$8,365,439	$8,602,124	$8,845,767	$9,096,575	$9,354,761	$9,620,546	$9,894,156	$10,175,824	$10,465,791
Debt Service	$(4,574,596)	$(4,574,596)	$(4,574,596)	$(4,574,596)	$(4,574,596)	$(4,574,596)	$(4,574,596)	$(4,574,596)	$(4,574,596)	$(4,574,596)	$(4,574,596)
Asset Management Fee	$(286,596)	$(286,596)	$(286,596)	$(286,596)	$(286,596)	$(286,596)	$(286,596)	$(286,596)	$(286,596)	$(286,596)	$(286,596)
	$-	$-	$-	$-	$-	$-	$-	$-	$-	$-	$-
Before Tax Cash Flow Available	$3,080,111	$3,274,318	$3,504,247	$3,740,932	$3,984,575	$4,235,382	$4,493,569	$4,759,354	$5,032,964	$5,314,632	$5,604,599

FIGURE 5.7   Sample 10-year pro forma.

(IRR), and equity multiples. Each of these is considered in the context of the required rate of return. The *required rate of return*, also known as the *hurdle rate*, is the minimum amount of interest, or *return*, that a developer is willing to accept on the money invested in particular project. This is often tied to the total interest rate the developer must earn in order to make payments on the borrowed capital needed for the project (debt and equity), also called the *cost of capital*. If the proposed project does not yield the required amount of return, it will not be considered a viable investment.

Note that developers use financial calculators or Excel formulas to calculate project returns quickly, but the actual equations are provided herein. This section is intended to introduce basic financial concepts but not explore them in depth.

Net Present Value. The only way to evaluate and compare earnings that are projected to happen two, five, 10, or more years into the future is to "translate" those projected future values into current values. Or, in other words, to determine what the annual income from a future year is worth in today's dollars. This is done through a process known as *discounting* in which pro forma generated future cash flows are revalued (or reduced) from a projected future value to a current value at a particular interest rate. Future values must always be discounted in order to arrive at present values because of a concept known as the *time value of money*, which assumes that the future value (or purchasing strength) of $1 is worth less than the current value of $1 due to inflationary forces. The resulting value is known as the *present value* of the future cash flow and is calculated on an annual basis as follows:

$$PV = \frac{CF_t}{(1+r)^n}$$

where PV = present value
CF = cash flow
t = future year
r = interest rate
n = total number of years

In isolation, the present value of an individual cash flow does not reveal much about the project's investment potential. However, when the sum of all project cash flows, including the anticipated residual from a future sale, are discounted together, they can be used to determine if the total cash flow received by the developer is greater than the developer's total initial investment in the project. This is known as the project's *net present value* (NPV) and can be calculated as follows:

$$NPV = \sum_{t=1}^{t} \frac{CF_t}{(1-r)^n} - PC$$

where PC = cost of the initial investment
CF = cash flow
t = future year
r = interest rate
n = total number of years

When the required rate of return is used for r to calculate the NPV, the developer will achieve his/her required rate when NPV = 0. This is the point at which revenue from all future cash flows (including the sale of the property) are exactly equal to the developer's investment cost at the necessary rate of return. Any positive value of NPV means that the project will yield a greater return than the required rate of return; any negative value of NPV means that the project failed to yield the required rate of return and should be deemed nonviable. Thus, a developer is willing to proceed with any project where NPV ≥ 0. Note that equity investors will calculate their own individual NPVs based on the amount of equity they contribute to the project and the returns paid to them on their investment.

Internal Rate of Return. NPV is not the only measure of return the developer will consider. A project's internal rate of return (IRR) is another important metric used to evaluate the attractiveness of the investment by considering cash inflows over time, including income upon sale, compared to the initial project cost. More specifically, IRR reveals the amount of return a project will generate, or, in other words, IRR is the interest rate at which the net present value of all the cash flows (both positive and negative) from a project equals zero. Mathematically, this can be achieved by setting the NPV equation equal to zero and solving for r. When the IRR of a project or an investment is greater than the minimum required rate of return, then the project or investment should be pursued.

Note that the value used for "cash flow" in the equation to calculate IRR can vary. Using BTCF reflects the use of debt equity while using NOI does not. Comparing these two versions of IRR allows the developer to evaluate the degree to which project returns are enhanced by leveraging the use of outside funds. Thus, these measures of returns are often referred to as the "levered" and "unlevered" IRR.

Equity Multiple. The equity multiple (EM) is a measure of how much money will be returned by a project for every $1 invested. The EM is used by equity investors as a quick method for comparing the returns of different projects. Equity investors expect project cash flows to return the amount of their original investment (return *of* equity) and also to generate payments above and beyond the amount of the original investment (return *on* equity). When calculating the EM, the total distribution of cash flows to investors includes both of these payments. It is a ratio of the amount of money invested by the equity partner as compared to the total amount of money returned to the equity partner, and can be calculated as:

$$EM = \frac{\text{total CF distributions recieved}}{\text{equity invested}}$$

Larger EM values are more desirable because they reflect a greater profit being returned to the equity partner for every $1 invested. However, note that the EM does not take into account when project returns occur and does not reflect the risk profile of the offering or any other variables potentially affecting the project's return. Although often used by investors, developers must also be aware of the EM potential of their projects in order to attract equity investment.

Public Sector Pro Formas. The public sector does not engage in developing projects for profit, which would seem to limit the need for a pro forma to model the financial outcome of an investment. However, as stewards of tax revenue and actors operating on a fixed annual budget without the benefit of traditional debt or equity infusions, the public sector *does* need to pay close attention to the financial aspects of its development projects: hard and soft project costs for construction, architect, and engineering services must still be tracked; annual operating costs for either building or infrastructure projects must be projected to ensure that operating funds will be available as needed; and "returns" may be reflected in the form of savings to the government from undertaking the project. Further, pro formas may allow agencies to evaluate alternatives and select one project over another.

However, public sector pro formas differ from those created by private sector developers. Perhaps the most noticeable difference is that, particularly for smaller projects, they often reflect all-cash transitions because government agencies must have funds available to cover costs before they are authorized to issue contracts and hire firms. Sources of funds may include appropriations from taxes, federal grants, or bond revenue, but will not include traditional debt financing or loan obligations. Larger infrastructure projects do sometimes rely on borrowed money, but the funds are borrowed internally from government sources at nominal or 0 percent interest rates. The closest government equivalent to debt financing is funding from federal or state "revolving funds" and state infrastructure banks, which require an agency to repay borrowed money in order that the funds might be used for another project. Additional public sector funding considerations will be discussed in Chap. 7.

5.3.3. Seeking Debt Financing

Most private sector development deals rely on sources of debt as part of their capital stack. Recall from Chap. 1 that a project's *capital stack* refers to the combination of funds from all sources of debt (from lenders) and equity (from equity partners and/or the developer). The debt portion of the capital stack will almost always be significantly larger than the equity portion. Thus, securing debt financing is a necessary milestone (Milestone 8) for developers in order to bring their projects to fruition. During the due diligence period, the developer will often work to secure a debt commitment from a lender in order to have certainty that the project can move forward.

Types of Loans. There are two loans that most developers must obtain in order to bring a project to fruition: construction loans and permanent loans. However, there are other types of loans, such as land loans and letters of credit, that are often involved in development projects. Note that lenders sometimes refer to loans as *liens* because they can be recorded on the property title. This section will provide details on several different loans types.

Land acquisition loans fund the acquisition or refinance of land. While developers typically use equity to buy project sites, doing so is not possible in every case. Thus, a land loan allows the developer to acquire a site when he/she may not have the necessary equity available. Land loans are generally used to fund the acquisition of greenfield sites; however, they can be used to acquire sites featuring existing improvements/buildings with the understanding that the site will be redeveloped and the developer is not using the land loan to buy and own an existing/operating property. To that end, land loans are generally only offered for a short period and a 3-year maximum loan term is common. The lender's expectation is that the loan is outstanding just long enough for the developer to complete entitlements and groundbreaking on a new project will occur soon thereafter. Land loans are considered to be one of the riskier loan types by banks because there is no underlying cash flow to support the ongoing debt service. Therefore, banks will attempt to mitigate their risk by requiring a 50 or 60 percent LTV and may also insist that the developer and not an LLC be personally liable for repaying the amount.

Construction loans, also known as *development loans*, are interest-only loans that fund the cost of developing a site. They may be provided in tandem with a land loan, in which case the bundled loan package is known as an *acquisition and development (A&D) loan*. Construction loans function similarly to a line of credit wherein the developer draws funds on an as-needed basis to cover the ongoing cost of site work and construction. The schedule by which the bank releases money to the developer under a construction loan is known as a *draw schedule*. This is often established with the bank as part of the loan approval process and may be tied to the completion of construction milestones. Draw schedules are important because they provide funds that allow developers to pay their contractors for labor, materials, and services during the actual construction of the building(s). The developer and attorney need to make sure that payment timing in any construction contract is coordinated with the bank's draw schedule. Because construction loans are interest-only accrual loans, the developer makes no debt service payments on the loan amount. Rather, interest on the total outstanding draw amount *accrues*, or accumulates, and is continually added to the total amount of the loan that must be repaid. Construction loans and all accrued interest are usually paid off by the placement of the permanent loan, although some construction loans include terms that allow them to convert to permanent loans.

Permanent loans are placed when the finished building reaches stabilization. Recall that stabilization is the point at which cash flow from rental income is sufficient to support operations and generate positive net revenues. Thus, in order for a permanent loan to be issued, construction must not only be completed but the resulting building(s) must be occupied by rent-paying tenants and operating contracts must be in place for management, services, etc. Permanent loans have conventional financing terms, meaning they are multi-year agreements that fully amortize by the end of the loan period. *Amortization* refers to the process of paying back the entire loan amount borrowed plus interest by the end of the loan term, usually 15, 20, or 30 years. Amortization happens naturally in conventional loans because the total debt service amount is made up of two different components: (1) an interest payment and (2) the repayment of some portion of the original amount borrowed. The schedule showing the calculations for each interest and principle payment amount as well as the outstanding loan balance is called an *amortization schedule*.

In addition to traditional loans, there are other forms of funding that developers often need to obtain. For example, in many states developers must post a bond to the local municipality in exchange for approvals and development permits. In the context of development, a *bond* is a form of financial guarantee offered by the developer to the municipality to ensure that any public infrastructure work associated with the project will be completed. Development bonds may also be referred to as *surety bonds, infrastructure bonds*, or *performance bonds*. Recall from Chap. 2 that the developer may be required to dedicate roads to the municipality; make off-site improvements to roads, sidewalks, or intersections; and may have to complete work related to water and sanitary sewer connections. The municipality must give a developer permission to begin construction before either the proposed project or the infrastructure work can be completed. However, once approvals are given, the municipality has no guarantee that the developer will actually complete the promised infrastructure work. A developer could fail to complete the necessary work for several reasons such as bankruptcy, deciding to sell the project before finishing it, changing plans, or by simply refusing to perform.

Rather than risk having to take a nonperforming developer to court, municipalities require the infrastructure work to be bonded so they have a means of completing it. Specifically, many infrastructure bonds are provided in the form of letters of credit. A *letter of credit* is a document issued by a bank on behalf of the developer that promises a third party the bank will cover certain expenses up to a given amount. Under qualifying circumstances, the third party can require payments be made by the bank up to the limit of the letter of credit. When used as a bond, the letter of credit is set equal to the cost of the infrastructure improvements and is issued to the benefit of the municipality. This allows the municipality to receive a payment from the bank to pay to complete the infrastructure work directly should the developer fail to perform. Assuming the developer completes the work as agreed, the municipality will release the bond upon inspection and acceptance of the infrastructure improvements. Note that getting bonds released can take several years and is not necessarily a priority for the municipality.

Lenders will usually provide an irrevocable letter of credit to satisfy the developer's bond requirement. By virtue of being irrevocable, the letter of credit is functionally similar to available cash from the municipality's perspective. Indeed, a letter of credit may be secured by the equivalent cash balance in the developer's bank account, which is then frozen until the bond is released. However, more often, the letter of credit is secured by a junior lien against the land with a high interest rate and an annual fee.

The Underwriting Process. In order to receive a bank loan, the developer must submit a comprehensive package of information to the bank containing specific details about the site, the proposed project, project costs, and a pro forma showing the financial profits expected to be generated by the operating asset after completion. This document is known as a *financing package* or a *loan package*. Gathering the necessary information can be challenging because the developer may not own the site nor have fully completed his/her due diligence before needing to seek a loan commitment in order to move forward with a project. Additionally, the developer must also submit proof that he/she has the financial capacity to undertake the project. In many cases, the need to demonstrate financial strength can lead the developer to seek partners, guarantors, or equity partners.

The developer's loan package will be examined by the lender through a process called *underwriting*, during which the lender will evaluate the project's overall risk and viability. Lenders will not necessarily rely on the developer's pro forma assumptions and may create their own financial models to test the viability of a project. They will also require verification of site conditions, such as site or environmental surveys, from the developer's team or third parties. At the end of the underwriting process, the lender will make a decision to approve or deny a loan.

If the lender believes that the project can be underwritten (or approved), the bank will proceed to negotiate loan terms with the developer. These include bank fees, the loan interest rate, repayment schedule, equity contributions, and other special terms. An *equity contribution* refers to a lender's requirement that the developer make a personal financial contribution to the project. Developers that do not have the financial wherewithal to meet the entire obligation, especially for large projects, may bring in guarantors or equity partners. Depending on the financial strength of the developer and his/her past performance, lenders may require the developer to be personally liable to the bank for losses should the project fail. Loans that include this requirement are known as a *personal recourse loans* while those that do not require it are known as *nonrecourse loans*. The recourse requirement is intended to incentivize the developer to be fully vested in the project's success. Loan recourse is more common for risky projects or those undertaken by small, entrepreneurial developers who may not have established a record of successful projects; however, banks can require recourse on any project.

The actual method by which banking lenders make funds available to developers is through debt instruments. A *debt instrument* provides evidence of a legally binding contract between the lender and the developer and contains the specific terms and conditions relevant to the loan and the project. Mortgages are common forms of debt instruments. In exchange for the loan, lenders will require the developer pledge something of value to protect the bank against loss in the event that the project fails and the developer defaults on the loan. This is called *collateralization*. In virtually all cases, the property itself is used as collateral and the bank's right to the property is captured by a *mortgage deed*, also called a *deed of trust*. Specifically, lenders require a "First Deed of Trust collateralized position," meaning that the bank is the first stakeholder to be repaid in the event of default. A First Deed of Trust is recorded on the title to the property itself and supersedes the claims of any of the developer's other investment partners. Lenders will also usually require the collateralization of development documents such as:

- Plans and specifications for the site and building

- General contractor's contract with the developer

- Architect and engineering services contract(s)

Traditional banks will hold the title to the property until the mortgage is paid in full.

5.3.4. Seeking Equity Investors

There are several reasons developers may seek equity investors. As discussed previously, lenders will not finance 100 percent of a development project, so the developer must contribute a certain amount of equity. However, there are significant growth and project limitations if the developer relies exclusively on his/her own capital to make up the project's equity contribution. Outside investment can propel a development project forward well beyond what is possible with internal investment alone while also allowing the developer to share project risk more broadly. Thus, developers will seek to raise equity contributions from outside investors. Recall from Chap. 1 that *equity partners* are private investors that lend money to developers in exchange for becoming limited partners in a development project and receiving preferred returns. Equity investors command higher returns than those paid on debt financing because, unlike banks, these investors do not necessarily have any claims on the property title and, thus, take a higher risk of not recapturing their investment if the project fails. Equity funds can be solicited at different phases of a development project, although they are most often sought once the developer is close to securing conventional debt financing or has a lending commitment and has certainty with respect to how much equity is needed to bring the project to fruition.

Just as there are many reasons for developers to seek equity partners, there are several reasons investors may be motivated to provide capital to developers. The appeal of lucrative returns and tax benefits are obvious incentives, but the opportunity to invest in a development project can offer significant value beyond financial returns. For example, acting as an equity partner gives investors the ability to participate in projects outside their primary skillset or knowledge base that they might otherwise not be able to capture. Investing in development projects also allows investors to diversify their overall portfolio beyond typical investments, such as stocks and bonds. Depending on the nature of the investment contract, direct investment in a development project may offers investors some degree of influence over major decisions, which is more than is available in stock-based equivalents like real estate investment trusts (REITs) or other similar investment vehicles.

Yet even with clear motives for both developers and investors to participate, seeking equity for a development project is anything but straightforward because every project is different and every investor has different objectives. It is the developer's task to align with investors when appropriate in order to build meaningful and continuing investment relationships. In other words, the developer's objective is not simply to raise capital, but raise the *right* capital for each project. Adding to the complexity of the undertaking, a developer's efforts to raise equity for a project are governed by the Securities Act of 1933 and may be subject to review by the Securities Exchange Commission (SEC), depending on the developer's approach to raising equity, the number of potential investors solicited, and their relative levels of wealth and sophistication.[6]

Despite the challenges, raising significant amounts of equity is certainly possible. Professional service firms can assist with raising capital, but many developers prefer to build and maintain their own investor network, often focusing on cultivating relationships before they actually need to raise money. Many developers begin raising equity within their current network of friends, family, and business partners. These are the people who already know and trust the developer. The importance of trust in an investment relationship cannot be overstated and, indeed, trust is often what compels people to invest in one development project over another. Over time, a developer can build his/her network by establishing a record of successful projects, earning a reputation as fair to investors, and by seeking referrals to new contacts. Referrals are often one of the most effective ways of leveraging the trust from a current network into new connections and developers are rarely hesitant about asking for introductions.

As a developer's investment network grows, it may become necessary to use customer relationship management (CRM) software to keep track of contacts as well as the specifics of each potential investor's criteria. The CRM can also help remind the developer how often to contact each potential investor. Along with contact information, the CRM document or system should specify:

- Investment threshold or amount of equity the investor is comfortable placing

- The type(s) of project that are of interest

- The goals of the investor

- Areas of concern for the investor

- Desired return

Recall that there are different financial measures of return that an investor may target. However, some investors may have a preference for other forms of return, such as projects that generate certain social or environmental outcomes. Understanding each investor's criteria helps the developer know what deals to present to them. It bears repeating that the developer's objective is not simply to raise capital, but raise the right capital for each project. On an individual basis, this helps the developer maintain and build the relationship by ensuring that an investor is comfortable with the risk, return, and goals of the proposed project.

When there is a match between the developer's proposed project and an investor (or group of investors), the developer must successfully communicate the details of the opportunity as well as the vision for the project, also known as the "development story," in order to raise equity. A clear and concise message drives the best outcomes in this regard. Specifically, the developer must be prepared to describe the core goals of the project and how decisions will be made to reach the desired outcome(s). The strength of the development team should also be addressed with the developer naming project team members, either key individuals or firms, and explaining how their individual competences will contribute to the project. While the project concept will remain the same, the developer can customize discussions for specific investors in order to demonstrate how the project can help them achieve their goals. The degree to which the story is compelling and satisfies a particular investor(s) needs, the greater the likelihood of raising the needed equity. In all cases, the developer should be clear about capital needs and projected returns, including the underlying assumptions.

Although conversations between the developer and individual investors represent a large portion of the developer's efforts to raise equity, certain information is also provided in the project's offering memorandum. An *offering memorandum* is a formal, written document that describes the proposed project and the investment opportunity for potential investors. Certain offering memoranda must be filed with the SEC and, with few exceptions, are made publicly available through the SEC's online database.[7] Developers should always consult with their attorneys for SEC compliance when creating a solicitation package and strategy. In general, offering memoranda should:

- Describe the developer's firm, past experience, and key team members.

- Describe the property and proposed project.

- Clearly outline sources of funding including debt financing, developer equity, and the total amount of investor equity being sought.

- Present clear pro forma assumptions and numbers.

- Explain under what conditions and when investors will receive payments.

- Explain what investor return will be and how it is calculated.

- Explain what will happen if revenue is not available to make a payment to investors.

- Convey both potential gains as well as risks.

5.4. SITE ACQUISITION

The many exercises and studies completed during the due diligence period are intended to provide the developer with information necessary to make a determination about the feasibility of the proposed project. Before the end of the contractually agreed due diligence period, the developer will use this information and make a final decision to either purchase or abandon the subject site. If, however, due diligence findings suggest a legitimate need for further research, such as the need to perform a Phase II ESA, the developer and seller may agree to an extension of the study period. However, it is possible that no extension will be granted, in which case the developer must make the acquisition decision based on the information available.

Once the decision has been made to move forward with the site purchase, there are still legal formalities that must be completed before the developer can finally achieve Milestone 5 (acquisition). Usually the developer is required by contract to provide the seller with a formal notification of the intention to acquire the site in advance of the due diligence expiration date. A settlement date, also known as a *closing* date, will then be agreed upon unless it is already be specified in the contract. Note that under some contracts, the developer is not required to purchase the property until after receiving zoning approvals. In such cases, site acquisition may not take place for several months or even years after due diligence studies are completed. Also note that under a ground lease scenario, "acquisition" is considered to occur at the point the lease becomes irrevocable, which may

or may not be after a due diligence study period depending on contract terms.

Prior to settlement, the developer's attorney will confirm that all seller obligations have been met and all conditions precedent have been satisfied. This typically requires supporting documentation from the seller or other sources, which will need to be received and accepted prior to the settlement date. The settlement itself is usually facilitated by a tile company, which will provide the necessary transfer documents to both the developer and seller for signature. Funds for payment of the outstanding balance of the purchase price will also be released at settlement. Depending on the developer's capital stack structure, the payment may include developer equity, funds provided by equity partners, and/or either a lender-provided land acquisition loan or acquisition and development (A&D) loan.

As previously described in this chapter, the actual process of securing site ownership is achieved by the registration (or recordation) of a deed of title in favor of the developer. The deed showing evidence of the transfer and the new ownership must be filed with the local jurisdiction, usually at the County Recorder's Office, in order to be formally recognized and valid. After acquiring a site, some developers may wait for changes in the market or other economic and/or political forces before pursuing development. However, in most cases, it is more advantageous for the developer to move forward with the approvals process.

CHAPTER CONCLUSION

This chapter described the many different kinds of research that are undertaken by specialized consultants and development team members during the due diligence period. This includes important legal and contractual considerations, site-based due diligence focusing on both technical studies as well as preliminary design, and also building a pro forma for financial analysis. Collectively, this information allows the developer to make an informed decision about either acquiring a site or continuing with the site search process. Completing due diligence (Milestone 4) and acquiring a site (Milestone 5) are important milestones in the development process; only upon achieving these two milestones can the developer finally begin work on the development project itself. The next chapter will discuss the approval process as well as the construction and subsequent operation or sale of the resulting building(s). These tasks represent the final outstanding development milestones.

REFERENCES

1. Maira, T., (2016), "Understanding Real Estate Ventures", Lexis Practice Advisor Journal, published online at: https://www.lexisnexis.com/lexis-practice-advisor/the-journal/b/lpa/default.aspx (accessed February 5, 2019).

2. American Society for Testing and Materials (ASTM) E1527-13, Standard Practice for Environmental Site Assessments: Phase I Environmental Site Assessment Process, ASTM International, West Conshohocken, PA, 2013, more information available at http://www.astm.org/cgi-bin/resolver.cgi?E1527-13 (accessed November 4, 2018).

3. 33 CFR Part 328; 33 U.S. Code § 1251; Federal Water Pollution Control Act as amended through P.L. 107–303, November 27, 2002, also known as the "Clean Water Act," and supplemental and related laws.

4. U.S. Army Corps of Engineers Headquarters website, section under Missions—Civil Works—Regulatory Program and Permits, "Regulatory Program Frequently Asked Questions," available at http://www.usace.army.mil/Missions/Civil-Works/Regulatory-Program-and-Permits/Frequently-Asked-Questions/ (accessed March 11, 2018).

5. EPA website, section on Laws & Regulations, "Summary of the National Environmental Policy Act," available at https://www.epa.gov/laws-regulations/summary-national-environmental-policy-act (accessed January 20, 2018); EPA NEPA website, available at https://www.epa.gov/nepa/what-national-environmental-policy-act (accessed January 20, 2018).

6. Securities Act of 1933; Securities Exchange Act of 1934; U.S. Securities and Exchange Commission website, "The Laws That Govern the Securities Industry," available at https://www.sec.gov/answers/about-lawsshtml.html (accessed November 4, 2018).

7. U.S. Securities and Exchange Commission website, "The Laws That Govern the Securities Industry," available at https://www.sec.gov/answers/about-lawsshtml.html (accessed November 4, 2018).

CHAPTER 6

APPROVALS, CONSTRUCTION, AND COMPLETION

Having purchased a site, it is usually in the developer's best interest to begin construction quickly. While it is sometimes possible to collect rental income from interim uses on a site, such as in redevelopment scenarios with an existing building, the developer only begins to earn pro forma projected revenues after the proposed new project is built and occupied. Additionally, the developer incurs expenses even while the site is undeveloped. Examples of such costs include property taxes; utility payments, if applicable; insurance premiums for liability and/or other coverage; and interest payments to equity partners, if applicable. Collectively, the costs required to maintain a property until development has occurred are informally known as *carry costs*, or the costs of financially "carrying" the property without earning a return. Despite these costs, there are select instances in which the developer may deliberately choose to delay construction. Examples of this include waiting to acquire other parcels for an assemblage or in cases where market demand has shifted dramatically and has repercussions for the proposed project. Despite strategic decisions to delay construction, at some point the developer must either decide to resell the land or achieve the remaining development milestones in order to complete the project. The final development milestones include obtaining approvals (Milestone 6), finalizing design (Milestone 7), securing debt financing (Milestone 8), construction (Milestone 9), and occupancy and operations or sale (Milestone 10). Note that because development is not a strictly linear process, debt financing (Milestone 8) may have already been secured through commitments made at an earlier stage of the development process; however, the actual placement of construction and permanent loans will typically not yet have occurred. This chapter will discuss the efforts of the developer and the development team to achieve these critical final goals.

6.1. PRECONSTRUCTION

Before construction can begin, the developer and development team must complete the project design and obtain approvals from the relevant jurisdiction. To achieve this, the developer and design team will likely begin to work with a construction firm to refine costs and design decisions. Once project materials are ready, the team will assist the developer throughout the approval process.

6.1.1. Preconstruction Services

The firm that will actually oversee the construction of a proposed project is referred to by several terms, including "builder," "construction firm," and "general contractor" or "GC." For consistency, this chapter uses the acronym GC. A GC is hired by the developer to provide materials and labor for construction and to oversee construction activities, including those of subcontractors. In a traditional *design-bid-build* project delivery system, architecture and engineering design services are independent of construction services and each firm works under a separate contract for the developer. However, the GC often has a role in the project long before construction actually begins. Preconstruction services can be provided through the early engagement of the GC, typically before (or concurrent with) the schematic phase of design, to assist the developer and design team in meeting design, price, and constructability project objectives. While this can occur during the due diligence phase, it may also occur substantially later if a construction firm has not yet been identified as part of the development team.

Among other valuable preconstruction contributions, a GC can provide ongoing budget feedback as the project design evolves beyond the conceptual stage. Indeed, the developer and design team may alter the project design based on cost estimates and value engineering suggested by the GC as a result of the GC's expertise and relationships with subcontractors. Preconstruction relationships can also lead to an early engagement of subcontractors and potentially lead to expediting initial construction phases, such as demolition or site work.

Construction firms may provide select preconstruction services on a pro bono basis but this depends on current market demand and the degree of effort contractors are willing to expend. In these instances, feedback is usually limited to the use of in-house resources to check prices and create a budget with limited subcontractor input. Rather than rely on pro bono arrangements, formal preconstruction services agreements are often used in which the GC will charge 0.25 to 0.50 percent of estimated construction hard costs, depending on the expected length of engagement. The value gained from preconstruction services is generally worth this relatively minor cost as it can lead to a better informed and more economical design. The fee generally does not fully cover the GC's time invested and carries the expectation the firm will likely be hired to build the project at the completion of the preconstruction services.

Under a preconstruction services agreement, the GC typically offers the following services:

- Multiple rounds of construction cost estimates at different design stages including conceptual design, schematic design, and final design.

- Utilizing key trade subcontractors and in-house experience to develop pricing.

- Ongoing pricing feedback to guide design decisions; for example, a cost analysis of a concrete versus steel structure given current market conditions.

- Maintaining a log of potential design options and alternatives for cost savings and value engineering recommendations.

- Performing a constructability review on proposed design variations and providing feedback on project plans and details based on GC experience with past projects and subcontractor input.

- Maintaining a project schedule.

Note that preconstruction budgets are estimates reflecting a professional opinion of the project's likely final construction hard costs made using limited design information; the GC does not provide the developer with a guaranteed commitment to a maximum cost as part of preconstruction activities.

If the developer does not engage a GC for preconstruction services, he/she may attempt to complete an internal design review and budgeting exercise with assistance from the architect, engineer, and other team members. However, these consultants typically do not have access to the continuous market cost feedback that GCs enjoy due to the frequency of their bidding activities and close subcontractor relationships. In rare instances, a developer may not need preconstruction services. This occurs most often in cases where the developer is regularly engaged in creating a standardized product, such as the development of several identical fast food restaurant chain buildings. In such cases, the developer may have extensive project pricing knowledge and the franchise may permit only limited design variations.

It is important to note that preconstruction services are sometimes provided as part of a *design-build* project delivery system in which all design and construction services are integrated and delivered to the developer through the GC as the single point of contact. In such arrangements, a separate preconstruction services agreement is not necessary. Subsequent sections of this chapter will discuss different project delivery methods in greater detail.

6.1.2. Schematic Design and Value Engineering

Schematic design represents a refinement of the preliminary, or conceptual, design and normally occurs after the completion of due diligence. During the schematic design phase, each member of the design team will prepare more detailed plans for their respective work by performing various analyses and evaluations and by refining and validating previous assumptions. As schematic design progresses, the developer and GC will use this information to prepare construction cost estimates in order to ensure the proposed project is achievable as originally envisioned or, if not, to identify adjustments to the design that must be made to meet project budget constraints. Thus, an important part of the schematic design stage is determining the most economical approach to design and construction.

The process of economizing the design program is known as *value engineering*. The main purpose of value engineering is to identify cost-saving opportunities for the developer without significantly altering the developer's proposed project program as initially captured in the concept design. This effort usually involves several members of the development team, including the architect, civil engineers, public agency reviewers and planners, the project attorney, the construction manager, and the GC. During a typical value engineering review, for example, the team may determine that reconfiguring lots for a residential development project may reduce street lengths, which could save on utility requirements. Similarly, the attorney and engineer can provide information to the developer about laws and regulations that may impact the project and the strategy for development, such as stormwater management regulations or the cost of improvements that arise from site plan regulations. Performing a value engineering study allows design and cost issues to be reviewed and potential options to be identified prior to final design. Design adjustments made during this stage will have less impact to the project schedule than changes made later.

Schematic design and value engineering exercises are also critical for validating the developer's pro forma projections, which form the financial basis of the project's feasibility. Preliminary cost estimates of the proposed site improvements, including the permitting or regulatory fees associated with the development program can be refined at this stage. These preliminary estimates are important for the developer to review in order to ensure the project costs are in line with expectations. However, the developer should continue to include design and construction contingencies in the pro forma to account for additional costs that may not be clarified until after the final design is complete.

The final product of the schematic design exercise is known as a *preliminary plan*, which represents an approximately 30 percent completion of the final drawings needed for the project. Although more detailed than the concept design, schematic designs are interim layouts that will serve to support subsequent, final detailed design efforts. Public review at this phase of the design process is extremely important and it is a sound business practice to engage public decision makers. If there will be political or procedural obstacles to the proposed plan, it is best for them to surface early in the design process. The level of detail included in the schematic design is required for most entitlement reviews and the developer will generally use schematic preliminary plans to begin the approval process. It is important to note that the design of the proposed project is always subject to change, particularly when it is presented to citizens and/or public review agencies for comment. Hence, the schematic design process is often an evolutionary and iterative process that can last for months or even years as the approval process progresses, particularly when rezoning for entitlement is required.

6.1.3. Approvals

As discussed in Chap. 2, development does not occur in isolation. Although the developer has been heavily focused on internal details and feasibility, the proposed project will inevitably impact the surrounding community and infrastructure systems. It is the role of local jurisdictions to administer a balance between the public interest and developer's private rights to developer his/her property. This is achieved through an approval process in which the proposed project is reviewed by relevant authorities. The purpose of the review is to evaluate the proposed project with respect to compliance with zoning ordinances, consistency with the current comprehensive plan, and evaluate land use impacts and areas requiring mitigation. The local government structure for land use approvals typically involves a zoning administrator, the Planning Commission, the Board of Zoning Appeals, and local governing bodies, such as the city council, town council, and/or county board of supervisors.

Different development projects require different approvals, depending on the subject site and the nature of the proposed project. For example, if the existing zoning does not allow the proposed project to be built by right, then the developer may need to seek a variance, change in zoning, or even an amendment to the comprehensive plan. For projects that do not require rezoning, the approvals related to those processes are obviously not required. However, all projects must obtain site plan approval and all necessary building permits before construction can begin. The basic process is essentially the same regardless of the type of approval needed, although certain approvals will be more lengthy and complicated than others. In most jurisdictions, the approval process involves the following steps:

1. Submission of project materials to planning staff.

2. Distribution of materials by planning staff to "review agencies."

3. Development team receives comments from review agencies.

4. Multiple rounds of receiving/responding to comments/re-review.

5. Effort to mitigate negative impacts.

6. Upon conclusion of staff comments and responses, the project is scheduled for public hearings.

7. Public hearing is held; public has the chance to comment.

8. Final staff recommendation of approval or denial.

9. If legislative action is required, legislative body approval or denial.

Courts accord great deference to the land use decisions of local governing bodies, although approvals can become overly politicized and/or unnecessarily restrictive. The developer will rely on his/her attorney's knowledge of legal limits of local government authority during the approval process. The overreach of public authorities often poses a legitimate concern for developers as well as a barrier to new development, which, despite some perceptions, can harm a community. In extreme cases, a developer/landowner aggrieved by a land use decision may have a constitutional claim under federal and state "takings" law if the land use regulation goes too far.

Several team members are involved in the approval process, including the architect, civil engineer, attorney, and broker. Each may assist in the preparation of submission materials as well as provide responses to comments and discussions with staff throughout the approval process. Note that if the developer is working with a build-to-suit client, the client should potentially also be involved in the approval process if doing so can help demonstrate the benefits of the project to the community. The following sections will discuss elements of the approval process in greater detail.

Community Involvement. Many jurisdictions promote community outreach and make notification of public parties a statutory requirement prior to, or immediately following, the submission of materials. Notification groups might include adjoining property owners, municipalities, community associations, and other civic groups. In addition to written notification, some jurisdictions also require that signs be posted on a property under review, stating the project name, type of project, case number, and a contact phone number. These measures are meant to ensure that the developer has at least minimal interaction with the community and that neighbors have the necessary information to become more informed or involved in the process if they so desire.

Some jurisdictions may actually require the development team to meet with citizens as part of the plan submission process. In these cases, notices of plan submissions are sent to interested civic groups, which are then invited to review the plans and discuss their concerns with staff. Citizen comments are often afforded significant weight and can influence the review staff, particularly in a rezoning case. These same groups may subsequently attend public hearings to place their comments and concerns on the official record. Although jurisdictions may see this process as being responsive to their citizens, it can also serve to provide a platform for obstructionism and "Not in My Backyard" (NIMBY) delaying tactics by those opposed to development. Another downside to placing an inordinate weight on citizen concerns is that they are often biased because each group usually only considers the development from its own perspective and, in most cases, the average citizen is not fully informed about zoning regulations, the development process, market conditions, or what is reasonably possible for a developer to support. When not managed productively, citizen opposition can become a major obstacle and result in delayed or denied project approval.

Therefore, to the degree possible, it is often in the developer's best interest to actively engage the community during the design process both before and during submission, regardless of whether such outreach is required. This can be critical for large, complex projects that will have a material impact on the surrounding area. For especially sensitive projects, the developer may hire special community consultants or public relations experts. Outreach allows the development team to consider the potential impacts on adjacent property owners, attempt to reconcile them with the developer's project requirements, and resolve potential conflicts with neighboring property owners early in the process if possible. It may be possible to help ease opposition by careful consideration of the road systems; attention to environmental issues, particularly tree preservation; and additional screening, landscaping, setbacks or other aesthetic devices. In the best scenario, the development team can potentially adjust project design to better fit into the community while also gaining support for the project. To the degree that plans reflect citizen concerns, the approval process may be smoother and the project more readily accepted during the public hearing phase. Further, many jurisdictions have an appeal period built into the approval process; developing a good relationship with community stakeholders may diminish the likelihood someone will appeal the decision of the approving body.

As part of a community outreach effort, design exercises may overlap with early marketing exercises in which the promotional team seeks to create a project identity for marketing. This may include a distinctive project name, logo, and color scheme that contribute to project appeal and vision. While not necessarily part of the technical drawings, these exercises incorporate design drawings to communicate a coherent project theme. Such materials can help with community outreach, approval hearings, and also when the developer eventually begins a marketing campaign to sell or lease the newly constructed asset. Note that the developer will most likely need to spend additional funds to for the creation of these materials and to hire consultants for an outreach and marketing campaign. Depending on the nature of the proposed project, the promotional team may include a commercial broker, public relations and/or marketing firm, and property management firm. Through their specialized experience, these team members can help the developer communicate the benefits of the proposed project both to the community and future potential tenants or buyers.

Submission, Review, and Revisions. Zoning ordinances and administrative regulations govern the submittal process in each jurisdiction. These sources of guidance will outline the process itself as well as the number and type of documents that need to be submitted, depending on the type of approval sought. The requirements for submission are usually extensive and extremely specific with respect to the information the developer must provide. Generally, submission materials must be received several months before any hearing can occur in order to allow time for the review process and the developer must account for this in project timeline estimates.

Submission Materials.

Many jurisdictions have established submission "checkpoints," depending on the nature of the approval sought. For example, an entitlement review requires a preliminary plan while a final site plan is required for final review, approval, and permitting. These checkpoints generally correspond to the land development design stages of conceptual, schematic, and final design efforts. Submitting plans and receiving feedback during each design phase creates additional work, but also allows the development team to gauge the success of their interpretations of local policies, regulations, and standards. This offers the developer incremental assurances that the project will be approved before expending significant time and money for the preparation of final layouts and detailed plans.

The submission of project design materials serves two purposes: (1) to communicate project details to reviewers and (2) to facilitate discussions. Rezoning, special use approvals, and special exception uses all rely on good graphical exhibits to convey information to those in the review process. The various site and building elements must adhere to the regulations of the local and state agencies and design documents indicate the project's compliance with these requirements. This includes representations of compliance with lot size requirements, bulk regulations, open space requirements, parking, signage, and landscape/screening. Plans also communicate how the project interacts with off-site elements, such as traffic, mass transit, stormwater management, and other community features. Sufficient detail is required to give all the agencies involved in the review enough information to evaluate the design from their respective disciplines. The submission of graphics and exhibits can also facilitate discussions between the development team and regulatory agencies. For example, if the developer is seeking a rezoning of the site, initial design concepts must be developed to a level that will allow discussion of alternative scenarios.

Review and Revisions.

The agency taking the lead role in performing plan review varies from jurisdiction to jurisdiction. In larger communities, the Office of Planning and Zoning (or equivalent) is given this responsibility; in smaller communities, an individual, such as the town manager, may be responsible for the review process. In some instances, a multipurpose agency established to administer building and environmental regulations may be created for this purpose. In whichever form, the designated lead review agency is also generally responsible for coordinating the review of plans by other governmental and quasi-governmental bodies and will distribute copies of the submitted materials to other agencies for review. Many of the agencies involved in early design reviews are also involved in final design review. The typical agencies involved in the review process include:

- Office of Planning and Zoning
- Health Department
- Engineering and public works
- Department of Transportation (DOT)
- Fire and safety protection
- Parks and recreation
- Building
- Utilities
- Stormwater (if separate)
- Soil conservation (erosion and sediment control)
- U.S. Army Corps of Engineers (for federal review of wetlands)

In some municipalities, courtesy copies of a submission are sent to elected or appointed officials who represent the area in which the project is located. Occasionally, plans may be delivered to adjoining municipalities in areas where regional planning efforts are impacted.

The number of required reviews varies by jurisdiction and plan complexity. In general, the early reviews will likely take longer. Some jurisdictions can review plans within a few weeks, while others may require several months to complete each review. As each agency reviews the developer's plans, they will either approve the submission or issue comments that require plan revisions. The developer can address these comments either by revising the design and changing the drawings or initiating discussions with the reviewer. In discussions, consultants and members of the development team must be prepared to defend their work technically, yet be flexible enough to accommodate modifications requested by review staff at each level of review.

Requirements and modifications introduced during the review process can materially affect the design and development program of the project. This often translates to an increase in project cost, not only because the developer must pay additional fees to the design team for rework, but because of the nature of the required changes themselves. For example, the developer may be asked to make changes that result in a less efficient site layout and lead to the development of less rentable space, may be required to pay higher impact fees than budgeted in the pro forma, or may have to incorporate additional elements as concessions to the community that were not included in the original project cost estimates. Such changes have implications for the pro forma and the financial returns of the project. In extreme cases, this can increase project costs to the point where the developer is no longer able to achieve the required rate of return and the project becomes nonviable. In such cases, review comments may be fiercely opposed by the developer. In contentious situations, the project attorney can support their developer-client by ensuring that requested changes do not overstep jurisdictional authority.

Regardless of the developer's preference, as the project moves through the approval process, project elements may need to be modified, relocated, or substituted to comply with the regulatory interpretations of reviewing agencies. This is less common with simple, by right projects, but more likely to be a challenge for large, novel, or complex projects. In the case of a development master plan, changes may focus on small, targeted components or dramatic sweeping alterations.

It is possible that financial or market hurdles will arise during a lengthy approvals process, which may require reconfiguration of project elements and phasing. For example, in assemblage projects the acquisition of needed parcels may be delayed or impossible, causing alternative design scenarios to be developed as a substitute. Similarly, market changes can occur that influence the mix of development elements such that, in a mixed-use development for example, the overall percentage of each product type (office, residential, retail) may be increased or decreased depending on shifting market demand.

After the land development team has addressed the comments by all relevant agencies and made any necessary design revisions, the plans are resubmitted for another review. The time necessary to make changes and resubmit new designs can be extensive. The resubmission gives reviewers assurances that their initial comments have been adequately addressed and changes properly integrated into the original design. When all of the review agencies have had their comments addressed, the plans are authorized for the next phase of the process: the public hearing.

Public Hearing, Approval, and Final Design.

A public hearing is the formal way in which municipalities conduct their official evaluation of a proposed project and render a decision to approve or deny it. Before the public hearing, the development team will have already met extensively with different review agencies and have worked to incorporate recommendations into the proposed project in order to improve the chances of approval. However, approval should never be assumed as given; unexpected challenges and opposition can occur during the hearing.

One of the greatest challenges of public hearings is opposition from citizens that have not been involved in the process but seek to use the hearing as an opportunity to further individual (or group) agendas. Unfortunately, rather than being informed members of the community working to make productive suggestions, some citizens are simply obstructionists. As mentioned previously, such individuals (or groups) will oppose any development, regardless of its value to the economy or the community, simply because they are opposed to change. Again, as previously stated, citizens are often biased and, in most cases, not fully informed about zoning regulations, market conditions, the development process, or what is reasonably possible for a developer to support. Citizen opposition can be a major obstacle, especially if it influences the results of the public hearing. Of course, not all citizen suggestions are uninformed or maliciously intended and legitimate concerns should never be ignored. Until further advancements are made in the use of technology, public hearings are the way jurisdictions give a voice to the community and ensure that concerns are addressed; thus, developers must be thoroughly prepared to engage with stakeholders in such formats.

After the hearing, the Planning Commission will make a formal recommendation with respect to the approval being sought for the proposed project. The governing body will then vote to either approve or deny the request as presented. If approval is denied, the developer may have to return to the design stage or may have the ability to appeal, as through the Board of Zoning Appeals for a rezoning application. In many jurisdictions, once a rezoning request is denied, the same plan cannot be resubmitted for at least 12 months. If the project is approved (Milestone 6), the development team can move forward to finalize project plans.

It is important to note that many jurisdictions place a limit on the length of time that an approval remains valid. This is done to prevent development projects from spanning across multiple years, which may render them exempt from compliance with the issuance of future changes to relevant laws and ordinances. Another motive for the expiration of development approvals is to enable a jurisdiction to more accurately monitor its rate and patterns of development. If approvals did not lapse, it would be extremely difficult to make reasonable estimations of growth and local governments would be unable to properly budget for the provision of public services and facilities.

Final Design. Completing the project's final design (Milestone 7) is a collaborative effort in which several members of the development team will be actively involved (Fig. 6.1). Final design efforts work toward producing a final site plan that is used for permitting and construction. The final design plan is commonly referred to as the *site plan* or *subdivision plan*. This phase is critical for improving plan clarity and facilitating the smooth construction of the project. The final design effort focuses on the technical details of the project for the site plan and related permit application documents. Each infrastructure system (roads, grading, stormwater, utilities, etc.) must be accounted for, coordinated, and designed with respect to the requirements of the developer and the jurisdiction. Sufficient information in the form of drawings, computations, details, narratives, and specifications must be provided such that regulatory agencies can review and approve the final site plan and contractors can actually build the project. Major changes to the designs or requirements at this stage will require significant rework and often result in project delays.

Some of the additional design documents that are often incorporated into the final site plan include:

- *Photometric plan:* A photometric plan, or illumination plan, is developed by a qualified professional that identifies the fixtures for site lighting along with a photometric analysis of the lighting levels.

- *Landscape plan:* The landscape plan is often included in the site plan, but prepared by a landscape architect. The landscape plan identifies proposed vegetation and details in hardscape features (patios and plaza) and site furnishings (seating, rails, stair finishes, etc.). This plan may also include tables of vegetation quantities, cover calculations, and preservation plans.

- *Architectural and building plans:* These plans are prepared by the building team (architect, mechanical, electrical, plumbing,

FIGURE 6.1   Team collaboration on final design.

and structural engineer) and often have their own specifications applicable to the project. The designs will be exchanged regularly with the site engineer to ensure the building details are coordinated (grading, door locations, footprint, utility connections, etc.).

- *Traffic studies and plans:* These project documents include additional information pertaining to the maintenance of traffic during construction operations and studies used for the basis of proposed road improvements.

- *Signage plans:* Specialty site signs (as opposed to traffic signs) are often developed for retail centers, campus settings, and residential communities. These designs are often customized and developed later in the design phase (often by architects or manufacturers).

- *Geotechnical plan and report:* While a site plan will generally include some geotechnical information to support infrastructure design (such as pavement, walls, building foundations, and slope requirements), a geotechnical report is often authored separately from the design documents.

- *Utility reports:* In addition to the computations provided with a site plan document, a drainage report may be required to provide additional detail on proposed utility designs that are shown in the site plan. Similarly, computations for water and sewer service may be required.

- *Survey plat:* The survey plat (or plats) focus on the proposed boundary and easement conditions for the project site (and possibly adjacent sites). The easement and boundary lines depicted in the plat should match the site plan information, but the plat is formatted for recordation purposes.

- *Specifications:* Some projects may reference standard specification from a local DOT or jurisdiction, but in many cases a set of specifications is developed and packaged separately from the drawings.

- *Supplemental site design:* Additional plan sheets may be developed by the design team to provide additional information to a contractor or owner. This may include plans for a temporary sales office or construction trailer site.

- *Stormwater management plan:* This includes runoff calculations to support quality and quantity solutions to the project's stormwater.

- *Erosion and sediment control plan:* These documents outline plans to control sediment during construction.

When necessary, each of these documents may be harmonized and combined into a single, coherent bid and/or permit document. Note that the developer must have approved final site plans and specifications in order to obtain certain building permits, without which construction cannot begin. In many jurisdictions, approved construction documents are part of the final plan; in others, the construction documents are a separate phase of the development process, not associated with the final plan.

Permits, Proffers, Fees, and Bonds.

As alluded to in the previous section, development approvals alone do not allow a project to begin. The final site plan, once approved by the local jurisdiction, is used to acquire permits and prepare the construction documents. No construction or other work can start until the relevant permits have been issued. Project timeline and costs can be significantly impacted if the permitting process is not incorporated into a developer's budget and schedule. Jurisdictions require the developer to obtain permits for each phase of the actual site development. Permits are typically required for site clearing and grading, erosion and sediment

control, demolition, and building/construction. The majority of the required permits are related directly to construction activity, building components, and utilities. However, other permits are also required. For example, an environmental permit is issued by federal, state, or local agencies for regulated activities to ensure compliance with environmental laws. Generally, it is either the developer's or GC's responsibility to secure permits, although the site civil engineer may be called upon for assistance. Members of the development team need to communicate clearly with respect to responsibility for permit action.

Permits serve to alert a municipality that construction is about to commence so that the necessary inspections can be instituted. They also serve as a review mechanism to ensure that construction is being performed with plans that are still current. Thus, as with most elements of development, permits are not issued without a review of the project plans. As with other approval procedures, the designated lead agency receives the final plans associated with a permit application and distributes them as necessary. Permit review is usually conducted in a single phase, although various components of the structure may be reviewed by different departments or agencies.

The primary focus of building permit review is on architectural and structural design, accessibility (where required), and connection to existing streets and other public infrastructure facilities. Plans are reviewed for compliance with appropriate municipal zoning requirements; local, state, and federal building codes; and other applicable regional and municipal regulations such as certain energy efficiency standards that have been codified or green building certification criteria that the municipality has adopted. With respect to zoning, the primary concerns are proper setback from property lines, area of building coverage, density and floor area, height, parking and other required facilities that support or are accessory to the principal use of the property. Determining compliance with building codes requires extensive analysis of construction details to ensure structural integrity and review of internal systems for heating, plumbing, and electric service. Building spaces, both interior and exterior, are assessed in terms of the accessibility guidelines. Fire protection systems and fire code construction standards are also reviewed.

An approved building permit and payment of associated fees give the owner/developer the legal authorization to start the construction of a building project in accordance with approved drawings and provisions. Generally, a building permit is required to erect, install, extend, alter, or repair a building and must be posted in a conspicuous place until the job is completed and passed as satisfactory by a municipal building inspector. For more detailed information on permits and permit applications, refer to the *Land Development Handbook, Fourth Edition.*

Fees and Proffers. In addition to the payment of application fees, jurisdictions also collect many other fees at the time of permit application, such as impact fees and/or connection charges as described in Chap. 2. Fee calculations are typically prescriptive and may be based on floor area, number of fixtures, projected usage, or other indicators. In some communities, impact fees, further proffers, or other development contributions to public services are also collected. This may include contributions to public schools, parks, or public facilities. *Proffers* are, in theory, voluntary contributions made by developers to the local jurisdiction to help offset costs associated with the impact of the development. However, in practice, proffer systems introduce ambiguity and the possibility that the developer will be asked for inequitable payments in order to secure approvals. Additionally, some jurisdictions also require the developer pay capital facilities charges.

Performance Bonds. Upon completion of the review and approval of the final site plan, additional administrative steps are necessary to ensure that any required public and/or infrastructure improvements are completed and conform to the approved plans. This is achieved through a developer guarantee made to the jurisdiction in the form of a performance agreement and bond (or other form of approved security), as described in Chap. 5. In most cases, these documents must be processed before the issuance of permits and construction commencement.

The bond agreement is a legal, binding contract between the developer and the permit issuing authority. It specifies the manner and date by which the construction of the public improvements and/or infrastructure, as shown on the approved plans, will be completed. The *performance bond* is a financial guarantee that covers the cost of the public improvements and/or infrastructure. Performance bonds are backed by developer funds and issued by a third-party financial institution, often the developer's lender or insurance company, with the jurisdiction as beneficiary. In the event that the developer abandons the project or otherwise fails to make the agreed public improvements, the financial institution will release the bonded funds to the local government. This allows the municipality to pay for the construction of the improvements, even if the developer is unable to complete them. Often performance bond guarantees are set at 110 percent of the construction costs to ensure that both construction costs and municipal oversight expenses are covered in the event the developer defaults. In this way, bonds guarantee construction of public improvements in accordance with approved plans.

Institutions that issue performance bonds usually examine the credit worthiness and past performance of the developer and can charge substantial fees for their service and the risk assumed. Like letters of credit, bond instruments have the effect of committing the developer's capital to the infrastructure project and funds may not be available to the developer for future ventures until the work is completed. This provides strong motivation to complete a project and fulfill the infrastructure obligations.

Performance bonds are generally not released until after a project is completed, occupied, or accepted by public authorities. This ensures that public infrastructure will be finished, maintained, and, if damaged, repaired by the developer. Many municipalities allow for a reduction or replacement of the performance bond when all wet utilities, grading, and paving is installed. In this case, a new maintenance bond is issued. Maintenance bonds are used to ensure quality of workmanship and cover repair costs if a developer defaults after the project is substantially complete.

6.1.4. Marketing for Sale or Lease

The marketing campaign to either sell or lease the project, depending on the developer's preferences, will often begin before construction starts. This is typical of leased projects in which lenders require the developer to demonstrate the project's viability by securing lease commitments from tenants prior to construction (pre-leasing). The project identity created during the design process may now become more formally integrated into the developer's marketing efforts, which may require a name, logo, images, and floor plans to be used in advertisements, flyers, a website, and other marketing materials. Thus, while the occupation and operation or sale of a completed project represents the final stage of a development project (Milestone 10), the success of achieving this milestone often depends on a marketing campaign that is initiated well in advance of the building's delivery.

For office, retail, industrial projects, a specialized commercial broker will be hired to represent the developer and solicit tenant interest. The broker will be responsible for qualifying potential tenants or buyers and negotiating to achieve the best terms possible on behalf of their developer-client, who is now referred to as the "landlord" for purposes of leasing arrangements or the "seller" in the event of a sale. Even if the developer intends to sell the building, it may still be necessary to go through the leasing process in order to create an attractive rental revenue stream for an investor. The period of time required to lease all (or most) available space is often referred to as the "lease-up" period and can vary depending on the project and market conditions. Recall from Chap. 3 that developers often prefer to lease large, contiguous spaces in new buildings and may make a strategic decision not to lease to smaller tenants for this reason, even if there is space available. As potential tenants are identified, the developer/landlord will evaluate each tenant's financial strength and ability to meet their rental obligations over the period of the lease, which may be as long as 10 years with additional options to renew. The developer/landlord's attorney will help prepare and negotiate lease and/or sales contracts as necessary. Note that in build-to-suit scenarios, as per Chap. 1, the developer may have designed the building for a particular buyer and already have an agreed sales contract in place for the property.

For single-family, multifamily rental, and condominium projects, leasing or sales agents will be retained. Most residential projects feature a dedicated sales or leasing suite that is located near the site during construction and will be moved on-site once the property can be occupied. The use of social media to generate interest in new projects is increasing for all types of development, but tends to be more prevalent for residential projects.

The cost of marketing campaigns can vary depending on project type and scope. Marketing for flex, industrial, and smaller in-fill projects is often less extensive, and therefore also often less expensive, than campaigns for high-profile mixed-use, office, and residential projects. Although marketing costs increase overall project costs, funds spent on marketing are necessary because generating interest in a project at this stage is critical. The developer has spent considerable time and money on due diligence, site acquisition, and approvals; there will be serious financial consequences if these costs cannot be recovered (through leasing or sale of the property) because market demand shifts or does not prove as robust as anticipated. Even the best marketing campaign cannot offset risk if the developer failed to properly account for the real estate market cycle. Without tenants or buyers, a project may be delayed or have to be reenvisioned, which may require new design and approvals. Even in speculative projects, where the developer is able to begin construction without significant leases or a sales contract, revenue must eventually be achieved or else the investment results in an empty building. In worst-case scenarios, developers may be forced to sell the land (or building) and absorb the financial loss, face foreclosure, or declare bankruptcy.

6.2. CONSTRUCTION

Despite achieving approvals and benefiting from a marketing campaign, no development project can be occupied unless it is actually built. Construction (Milestone 9) is a critical, and often the most visible, milestone in the development process. The construction process is different for residential developers of single-family homes than it is for other commercial developers, including multi-family and multistory condominium developers. While a residential developer will sometimes build homes, he/she will often obtain approvals and then release lots to a homebuilder, who is responsible for the construction and sale of the homes. Additional lots are released by the developer after the preceding supply of homes is sold. Regardless of what entity does the building, single-family home construction is distinct from other forms of commercial construction. Similarities do exist, of course, including the need for compliance with jurisdictional inspections and applicable construction health and safety regulations. However, there are many key differences. For example, single-family homes can be completed relatively quickly and do not require tower cranes, extensive subterranean excavation, or the installation of complex building systems. The following sections will include information that applies to both homebuilding and commercial construction; however, focus will be predominately on the commercial sector.

6.2.1. Construction Documents, Project Delivery Methods, and Contracting

Before construction begins, construction documents must be completed and a CG must be hired. *Construction documents* include all written and graphic documents prepared to communicate the project design and construction requirements. This includes bidding/procurement requirements, contracting requirements, specifications, and contract drawings. Construction drawings may expand upon the final site plan to include additional, developer-requested details for the final design. These may include enhanced landscaping plans or value engineering decisions that go beyond the requirements for site plan approval from the local jurisdiction. *Specifications* provide a written description of the project requirements, construction materials, and the quality and performance standards for the products, systems, and equipment to be incorporated into the project. Specifications complement the drawings by providing an opportunity to communicate information to the

builder/contractor that cannot be fully provided through the drawings. In order to bid on a project in a traditional format, a GC will need this information in order to prepare an informed bid.

In the traditional delivery format, the developer first contracts with the design team to create project design and construction documents. These are then circulated to GCs to solicit bids to perform the construction work. Note that GCs will include a detailed list of *qualifications and assumptions* in their bids to clarify items that are included or excluded from their scope of work and pricing, as well as assumptions made where items are not shown or unclear; these should be incorporated into the construction contract. After reviewing bids, the developer enters into a contract with the selected GC for the project construction that is separate from the design contract. This process is often referred to as the *design-bid-build* delivery method and has historically been the most common method used for the development of commercial projects. However, other project delivery methods are increasingly being used by both private and public sector owners that integrate the design and construction project components, such as:

- *Design-build:* The developer contracts with a team that includes the design professionals and the GC for the project. The GC typically acts as the team lead. In design-build scenarios, it is possible for work to begin before final design is completed.

- *Value-based award:* The developer contracts with design professionals to prepare appropriate project construction documents. Bids are then solicited from GCs with requirements for separate price and technical proposals. The contract award is based on a weighted evaluation of technical qualifications and price.

- *Design-build-finance:* The developer contracts with a team that includes the design professionals, GC, and a financial institution. The team is under contract to deliver the project, including long-term project financing.

Contracts for each delivery method can be based on several different pricing structures. Two common examples are *firm fixed price*, also called *lump sum* contracts, and *guaranteed maximum price* (GMP) contracts, both of which are common for small- to medium-sized building projects. These contract forms are sometimes also known as "construction manager at-risk" because the party managing the construction bears most of the risk for cost overruns. Large complex projects, such as hospitals, stadiums, or multi-building sites, require various specialized contracts tying together several layers of management and consultants. Construction agreements are usually based on commonly used templates from the American Institute of Architects (AIA) or the Design-Build Institute of America (DBIA), but can also be customized.

In lump sum contracts, the GC hires subcontractors to perform various trade work related to concrete, mechanical, and drywall. The GC is responsible for managing and coordinating all project trades, but may perform select work as well. The GC is responsible to deliver the project as per construction plans at the agreed total price and carries the risk of cost overruns or pricing omissions. Conversely, the GC will benefit from any cost savings. It is important to note that the GC is responsible only for work indicated on contract documents and any additional work added after the contract award that warrants a change order will increase the contract sum. Schedule issues can create schedule related claims where the contractor may be due additional time and related compensation. These include weather delays beyond a contractually defined number of acceptable weather-loss workdays; unexpected site conditions, bad soil, or contamination; and design errors and missing plan details. Lump sum contracts are preferred by federal and other government agencies because they shift risk to the contractor and allow the agency to easily price-compare multiple competing bids.

Under GMP contracts, the GC's responsibility is similar as in the lump sum contracts described above, but a contingency line item is included in the total price to account for cost overruns or

schedule issues. The expectation is that minor issues are accounted for in the overall GMP and the contractor is not entitled to additional compensation for small items that may not be fully shown in the plan but which are implied or typically required. The developer, GC, and attorney need to ensure the contract is clear on the definition of "minor issues" to make sure that reasonable expectations are agreed upon. Additionally, GMP contracts include a "savings split" in which the developer and GC share in cost savings. This is often the preferred contract method for private developers because the ability to lock down construction pricing earlier can help with financing. GMP contracts also allow the developer to share in the cost risk/rewards with the GC.

6.2.2. Placing the Construction Loan

The developer will most likely need to obtain a construction loan in order to begin construction activities. Recall from Chap. 5 that construction loans are interest-only loans that function similarly to a line of credit wherein the developer draws funds on an as-needed basis to cover the ongoing cost of site work and construction. Draw schedules provide funds that allow developers to pay their contractors for labor, materials, and services during the actual construction of the building(s). If the developer has not previously secured a financing commitment for the project (Milestone 8), this will need to be completed in order for the project to continue. Before the GC is able to begin construction, the developer will need to secure construction financing and ensure that funds are available for disbursements when needed.

6.2.3. Construction Phases

After being awarded a construction contract, the GC will have key staff and subcontractors in-place to support the project, but will not begin until the developer issues a Notice to Proceed. A *Notice to Proceed* is a formal written notice to the contractor authorizing work to begin. Different notices to proceed could be used for different project phases, such as for site work, or one notice can be used for the entire project. Once permission has been granted for the work to begin, construction typically moves forward through the following phases:

- Site work
- Foundations
- Structure
- Interior framing
- Installation of mechanical, electrical, and plumbing (MEP) systems
- Life safety
- Building envelope
- Interior finishes
- Fixtures, finishes, and equipment (FF&E)

Different work phases usually require the skills of different specialized subcontractors and journeymen. As many as 25 to 50 different subcontractors may be included at different times throughout the construction process. Part of the GC's responsibility is to manage team schedules in order to maintain continuous progress and reduce delays.

Construction work related to preparing the site to receive a building is sometimes referred to as *horizontal construction* whereas work to construct the actual building itself is sometimes called *vertical construction*. The following provides an overview of each of the different construction phases, including both horizontal and vertical work.

Site work generally refers to the demolition of any existing structures and the subsequent preparation of the site, including earthwork (clearing, grubbing, rough and final grading), utilities work (wet and dry), landscaping, and hardscapes (sidewalks, pavers, etc.). Once site work is completed, the building's foundations can be installed. *Foundations* serve to transfer the gravity load from the building structure to the earth (soil, rock, etc.) underneath. Typical foundation systems include spread footings, mat slabs, geopiers, and caissons. The type of footing used will be informed by recommendations in the geotechnical reports because the type of soil and depth of necessary excavation are two predominant factors in selecting the foundation system. Based on consultations with the geotechnical engineer to evaluate design

options, a structural engineer will design appropriate foundations for the proposed building and site conditions. Note that decisions about foundations may have taken place relatively early in the design process, yet site conditions may subsequently be discovered during excavation that are materially different than expected based on previous due diligence. Examples of unexpected conditions include encountering rock or soil that has previously been excavated. These and other conditions may require changes to foundation design, which can impact the construction schedule and project costs.

The building *structure* refers to the frame of the building that supports the weight of non-structural elements and transfers all loads down to the foundations. Concrete, steel, engineered metal studs, and wood studs are all typical elements of structural systems. The structure must be completed before other systems can be installed. *Interior framing* refers to walls that are nonstructural and can be removed without needing to brace the remaining structure. Metal or wood studs are common framing materials, and aluminum framing is sometimes used for glass storefronts. *Mechanical, electrical, and plumbing* (MEP) refers to systems involved in heating, cooling, powering, lighting, and plumbing a building. Each of these different systems is critical for user experience in the finished building. Design choices with respect to MEP systems also impact overall project costs. Indeed, MEP systems can often comprise 25 to 50 percent of the overall project hard costs. The two main *life safety* systems featured in modern buildings are the fire alarm and sprinkler systems. Other subsystems, such as smoke evacuation or CO monitoring, may be related to or tie into the fire alarm system. The choice of placement of life safety systems is often code driven based on the use and construction type of the building. The *building envelope* refers to the entire exterior wall section. This is sometimes called the "skin" of the building and is usually thought of as the outside materials that are externally visible on the building, such as brick, glass, and siding. However, the building envelope also includes, air gaps, flashing, insulation, sheathing, framing, and other components. Envelope systems need to be carefully designed to prevent water/moisture infiltration.

FIGURE 6.2 Different stages of vertical construction.

Figure 6.2 shows a building structure before and after it is enclosed by the building envelope. Once the building is enclosed, the construction team can begin to install finishes, such as paint, carpet, and tile. Furniture, fixtures, and equipment (FF&E) and operating supplies and equipment (OS&E) can then also be installed. These are typically owner-provided items, such as furniture for the lobby of an apartment building (FF&E) or kitchen supplies for a hotel restaurant (OS&E).

The jurisdiction will carry out required periodic inspections of permitted work throughout the different phases of construction. These are typically related to structural, MEP, life safety, and accessibility systems. Jurisdictions are not the only entities scrutinizing work during construction. Third-party inspections and testing are often also required by either building codes or the engineer of record (or both). In a third-party inspection, an independent agency will test key elements of the project, such as structural concrete or mechanical/electrical systems. The various inspections much be coordinated with site activities to maintain the project schedule. Failure to satisfy an inspection, especially from the jurisdiction, can delay work and jeopardize building delivery. This can be especially problematic because occupancy obligations under preconstruction leases represent a legal obligation for the developer to provide tenant space as and when agreed.

6.2.4. Change Orders

It is rare that final site plans do not need changing once construction commences. Changing field conditions, discovery of previously unidentified information affecting design, late changes in the building design, and project adaptations due to shifting economic conditions are all possible causes of plan revisions. When revisions are needed, the submitting engineer must prepare a formal plan revision and submit the updated plans to the local jurisdiction for approval. Given that construction is usually underway in these instances, quick review and approval are essential. Most jurisdictions give priority treatment to revisions of already approved plans.

Changes to construction documents are captured by *change orders* to the GC. These can affect construction timeline and cost and can be additive (increasing cost/time) or deductive (reducing cost/time), depending on the nature of the situation. Some scenarios that can lead to change orders include:

- Differing site conditions/unforeseen conditions
- Weather delays/impacts
- Design bulletins/addenda affecting price or schedule
- Owner-driven changes
- Reconciliation of allowances or unit prices with final quantities

6.3. COMPLETION

The term "completion" can have several different meanings in the context of development. The two most common uses refer either to the completion of construction, as defined in various contracts, or to the completion of the development project itself. It is important for team members to be aware of the potential differing uses of the term to avoid confusion.

Engineers and construction professionals tend to think a project is "complete" at the end of their scope of work, which is largely connected to construction activities. Even through this lens, there can be ambiguity with respect to determining exactly when the construction portion of a project is actually completed. Some measures may tie it to contractually defined *substantial completion* conditions or to the receipt of a certificate of occupancy from the local government. In a construction context, there may also be considerable "punch list" items yet to be completed even after the building is "complete"; the developer typically holds back a portion of the construction contract fees, known as *retainage*, until such items are finished.

However, from the developer's perspective, a vacant building that generates no cash flows and continues to accrue interest on an outstanding construction loan does *not* represent a completed project. Once construction is finished, the developer must still reach Milestone 10, occupancy and operations or sale, in order for the project to be completed. The building must be leased or sold in order to create economic value and support the debt that was used to finance the project and pay development team members' fees. Thus, developers and brokers may often refer to the end of construction as *building delivery*. A building is said to have been *delivered* to the market when tenants can begin to take occupancy. Only at this point is the developer finally in a position to start recognizing cash flow from the development project. Further, even after the first tenants assume occupancy, it can still take time before the building reaches *stabilization*, the point at which cash flow from rental income is sufficient to support operations and generate net positive revenues. Therefore, it is important for the development team to recognize that their developer-client may not consider a project to have reached "completion" until stabilization and the placement of the permanent loan occur. It is possible that a developer may contractually define the end of the construction period as

FIGURE 6.3   Building ready for occupancy.

the date on which the construction loan is retired, which can more closely align construction completion and project completion (Fig. 6.3).

6.3.1. Construction Closeout, Final Inspection, and Certificate of Occupancy

Construction *closeout* involves a myriad of activities that occur once construction is completed. Systems will be tested to ensure they are functioning correctly. A range of final submittals must be made including operation and maintenance manuals, as-built drawings, and test reports. Final inspections must also be performed.

Most jurisdictions require the development team to produce record document plans, sometimes called *as-builts*, upon completing construction. *Record document plans* are drawings that show modifications to the project's construction drawings to reflect changes made during construction due to field conditions or other factors. They show the boundary of the site and final location of all buildings; horizontal and vertical location and size of pipes and apertures; location of fire hydrants; and location and width of streets, walks, trails, and other improvements as constructed. These drawings serve several important purposes.

First, they offer local jurisdictions some assurance that the actual construction was performed is in accordance with approved plans and revisions. Second, they present government agencies with the final opportunity to verify that construction complies with local regulations and standards. This is necessary to ensure that systems will function properly and not cause unanticipated impacts to the larger community. Finally, the plans facilitate future maintenance and repair, particularly for underground utilities whose location, sizing and method of construction are otherwise hidden from view.

Most municipalities require applications for occupancy permits prior to the actual, full-time occupancy of the premises. This permit stage allows the municipality to perform a final inspection of the structure to verify that it was built as required. A *certificate of occupancy*, sometimes referred to in short as a "C of O," is a document of authorization from the municipality allowing a newly constructed building to be inhabited. This document certifies the construction is in compliance with public health and safety codes and all applicable building codes. Receiving the C of O is important to the developer as it is typically required by the lender prior to closing on the permanent loan. The certificate will not be issued until all site and building improvements have been completed and documented.

Jurisdictions may also require the submission of postconstruction documents to confirm that buildings, including site infrastructure components, are structurally sound and built in conformance to approved plans and local, state, and federal applicable codes. Examples of post construction requirements in certain jurisdictions include:

- *Final location surveys* (also referred to as record document plans)—a drawing showing the location of the building and any other above grade (visible) improvements to the property. These improvements consist of driveways, sidewalks, fences, and retaining walls. The drawing must be signed and sealed by a licensed land surveyor. Typically, the lender will require the property corners to be set.

The accuracy level shown on these drawings varies according to local code but at a minimum, is equivalent to the approved plans.

- *Landscape certifications*—verifies the property has the correct number of trees, accurate in terms of size and type as compared to approved plans. It ensures the lender that proper stabilization measures including sod installation and other vegetative controls have been provided according to the approved plans and issued permits.

- *Elevations certifications*—verifies that the elevation of the first floor of the building is set within tolerance levels specified in the contract documents. Some municipalities require photography to be shot around the building to confirm the conditions in the field match the proposed condition on the site plan. This is particularly important when gauging accessibility compliance.

Finally, when the bonded public improvements and/or infrastructure components of the project are completed, the jurisdiction must inspect, approve, and accept them before the developer can turn responsibility for maintenance over to the public sector. For streets, this includes all final paving, signage and marking, street lighting, street side landscaping, and any utilities must be complete and correct per plan/permit; all broken, cracked, or otherwise damaged curbing, driveways, or other street components must be repaired or replaced. The developer's performance bond will not be released until all required work is completed and accepted. The process of getting a bond released can be arduous if the jurisdiction is slow to make inspections or is overly critical of conditions. It can take as long as 12 to 24 months after construction is complete for the developer to regain access to the funds held for the bond.

6.3.2. Occupancy and Operations or Sale

As stated previously, from the developer's perspective, a vacant building that generates no cash flows and continues to accrue interest on an outstanding construction loan does *not* represent a completed project. The final milestone in the development process (Milestone 10) is achieved once a newly built asset is either occupied and begins operations or is sold. Only at this point does the developer begin to recognize economic value from the development process. There are several final activities that must be completed in order for the developer to move successfully into this stage.

Tenant Build-Out.

In order to use their leased space, commercial tenants, particularly office and retail tenants, may require substantial changes to the "warm, lit shell" or "vanilla shell" provided by the developer. Recall from Chap. 3 that any changes or customization to this basic space are known as the tenant's "build-out," or may also be referred to as "fit-out." In most cases, this work must be completed before tenants can occupy and use the space. Further, investors may require a merchant builder to fully lease the building before acquiring the property. This allows the investor to buy an existing revenue stream of rental income without having to engage in interior construction. Tenant build-out may or may not be included as part of the GCs construction work, depending on the nature of the tenant and project.

For retail tenants, the developer may provide a period of free or reduced rent during which time the tenant can undertake their own build-out. This allows each retailer to use fixtures, interior layout, and other design elements to create a customized experience for its customers. This also allows restaurants to create their own kitchen space with such specialized equipment as may be necessary for their business. Retail tenants will try to obtain any necessary licenses, such as a liquor license, during the build-out period. In instances where a free rent period is established for a fixed period of time in the lease, it is in the tenant's best interest to have a build-out completed quickly, lest they be obligated to begin paying rent on a space they cannot yet use. Note that during this time, the tenant is said to have taken possession of the space, even if they are not physically occupying it on a daily basis. This is significant because any lease requirements for tenant insurance, utilities costs, or other obligations may come into effect based on possession rather than occupancy.

Unlike retail tenants, office tenants typically require the developer to build-out suite space on their behalf. Any rent-free periods typically begin after the build-out is complete and the tenant has taken occupancy of the space.

Turning over Operations to a Management firm or HOA.

As the developer transitions toward the role of landlord for a particular project, building operations will eventually be turned over to a management firm or homeowners' association (HOA). This allows the developer to refocus his/her energy on looking for new opportunities rather than managing an existing property. This shift may coincide with a certain level of leasing or, in the case of single-family home or condominium sales, may be tied to a particular percentage of units sold. Property management of offices, retail centers, multifamily apartments and other commercial buildings is a specialized sector of the real estate industry. Large development firms may have internal divisions that handle property management for their portfolio of assets. However, developers are equally likely to contract with a third-party firm for property management services. Management responsibilities, whether internal or contracted, include maintaining and repairing the building, collecting rent, addressing tenant concerns, security, paying property bills and taxes, maintaining landscaping, snow removal, and other similar activities.

It is not only building operations that the developer must transfer: many jurisdictions require certain components of a project's future operations and management to be specified on plans and plats. For example, a newly installed utility pole may belong to the electric company, the cable company, the telephone company, or an HOA; without a field indicator or accurate mapping, it may be a slow and inefficient process to determine future responsibility for the maintenance of the pole. Similar issues can arise with respect to future maintenance responsibilities for peripheral landscaping, utilities infrastructure, stormwater management systems, and streets. It is preferable for the developer and other parties to have clarity on their respective responsibilities than to inadvertently discover an oversight and possible damages in the future.

Stabilization and Placing the Permanent Loan.

When commercial projects, including multifamily apartment rentals, reach a level of occupancy where cash flow from rental income is sufficient to support ongoing operations and generate net positive revenues, the asset is said to have stabilized. It can take a full year (or perhaps more) to reach stabilization, depending on factors such as the overall economy, market demand, success of marketing efforts, and the desirability of the project. Stabilization is a critical point because the developer can finally retire the construction loan and place the permanent loan. Given that the permanent loan will remain on the property until it matures, is refinanced, or the asset is sold, this is the final component of the debt-financing milestone. Once the lender has verified the property's operating accounts and received any final assurances that may be required, the funds will be released. At this point, the developer has finally achieved all 10 development milestones and completed the project.

CHAPTER CONCLUSION

This chapter described the final milestones necessary to complete a development project. Obtaining approvals, finalizing design, and construction can be difficult and lengthy processes, yet the developer will not begin to earn revenue until each of these final goals has been reached. Further, previous commitments for debt financing are realized as the developer draws on construction loans and finally converts to a permanent financing structure upon stabilization of the asset. Once the finished project is ready for occupancy and operations, the developer's role transitions to that of a landlord or, in the case of a merchant builder, the asset is sold. While many of the development milestones are the same for both private sector and public sector developers, there are important differences in the public sector process. The next chapter will discuss the unique challenges of public sector development projects in greater detail.

CHAPTER 7

SPECIAL CONSIDERATIONS FOR PUBLIC SECTOR DEVELOPMENT

Thus far, this book has described the critical milestones of the development process in terms that generally apply to both private and public sector developers and has pointed out key differences where appropriate. This chapter is dedicated solely to public sector development projects at both the state and federal levels and their unique considerations. There are many similarities between federal and state systems for development projects: public sector development projects at both levels are subject to funding limitations, changes in political will, and formalized bidding processes. However, it is important to realize that they are not identical. Federal regulations and processes will be the same anywhere in the country, but at the state and local levels, though they may be conceptually similar, the specifics of law, policy, and process will vary.

7.1. KEY DIFFERENCES BETWEEN PUBLIC AND PRIVATE PROJECTS

The public sector is an extremely active developer, responsible for both infrastructure and building projects. In the public realm, development projects are often classified as government "procurements" and may also referred to as "acquisitions."[1] Procurement (or acquisition) activities of the federal government specifically include construction, engineering, and architectural services for real estate development.

The federal government delegates authority for development to different federal agencies, empowering them to solicit, award, and oversee design and construction services. In the Executive Branch, the largest sector of government, these actions are most often performed by a specialized agency called the General Services Administration (GSA). The GSA has two major divisions: one for the procurement of supplies and the other related to buildings and development projects. Certain Executive Branch agencies, such as the Department of Defense, have special authority to oversee their own development projects without GSA participation.

States may also have a state-level GSA (or sometimes Office of Procurement or other variation) or else will delegate authority to a specialized agency dedicated to handling state building and construction. For example, in Ohio the Ohio Facilities Construction Commission (OFCC) is empowered by the Ohio Revised Code:

> ...to contract for and have general supervision over the construction of any projects, improvements, or public buildings constructed for a state agency, to include the design, specifications, inspection, etc. of such construction projects. When a state agency wants to construct a new building or structure, or needs to make an alteration to an existing building or structure, it

may contact the OFCC for initial guidance and recommendations.[2]

Note that the OFCC's purview is limited to buildings. As is the case in most states, road and other transportation infrastructure projects are the responsibility of the state Department of Transportation, in this case Ohio Department of Transportation.[3]

7.1.1. Regulatory Environment

The processes of public sector developers at both the state and federal levels are heavily regulated. Though technically governed by different sources of law, there are many similarities between state and federal regulations because state legislation is often derived from federal codes. Further, when a state receives federal funding for a project, such as for National Highway System (NHS) road construction, the state must also satisfy the applicable federal regulations.

The three main governing sources of federal regulations are contained in:

- The U.S. Code (USC)
- The Code of Federal Regulations (CFR)
- The Federal Acquisition Regulations (FAR)

Each of these sources provides detailed definitions and instructions specifying how the federal government must undertake a procurement action. Wading fully into the labyrinth of federal procurement laws is beyond the purview of this book; however, a brief mapping exercise of some of the relevant clauses is useful for conveying the level of complexity involved in federal regulations. This is important to understand for GCs, civil engineers, and other firms aspiring to work with public sector developers.

Some of the most important titles of the USC with respect to federal public sector development projects are Title 23 (Highways), Title 40 (Public Buildings, Property, and Works), and Title 41 (Public Contracts). Figure 7.1 shows an abbreviated map highlighting specific sections of these titles; note that this is an abbreviated illustration and not an exhaustive representation of the regulatory environment.

It is important to understand how the clauses of these titles are interconnected and, in the case of Title 23, how they bind states to federal procedures. For example, 23 USC § 112(a) establishes that the state level department of transportation must comply with the federal Secretary of Transportation with respect to Federal-aid highway bids:

> In all cases where the construction is to be performed by the State transportation department or under its supervision, a request for submission of bids shall be made by advertisement unless some other method is approved by the Secretary. The Secretary shall require such plans and specifications and such methods of bidding as shall be effective in securing competition.

This is further established by instruction given to states in 23 USC § 112 (b)(1):

> Subject to paragraphs (2) and (3), construction of each project, subject to the provisions of subsection (a) of this section, shall be performed by contract awarded by competitive bidding, unless the State transportation department demonstrates, to the satisfaction of the Secretary, that some other method is more cost effective or that an emergency exists. Contracts for the construction of each project shall be awarded only on the basis of the lowest responsive bid submitted by a bidder meeting established criteria of responsibility. No requirement or obligation shall be imposed as a condition precedent to the award of a contract to such bidder for a project, or to the Secretary's concurrence in the award of a contract to such bidder, unless such requirement or obligation is otherwise lawful and is specifically set forth in the advertised specifications.

Note that the guidance provided in 23 USC § 112 (2)(A) effectively unites the processes for the award of technical contracts under Titles 23 and 40:

> Subject to paragraph (3), each contract for program management, construction management, feasibility studies, preliminary engineering, design, engineering, surveying, mapping, or architectural related services with respect to a project subject to the provisions of subsection (a) of this section shall be awarded in the same manner as a contract for architectural and engineering services is negotiated under chapter 11 of title 40.

Finally, 23 USC § 112 (2)(B) makes reference to the need for compliance with the FAR and CFR:

> Any contract or subcontract awarded in accordance with subparagraph (A), whether funded in whole or in part with Federal-aid highway funds, shall be performed and audited in compliance

United States Code
    Title 23: Highways
        Chapter 1 - Federal-aid Highways
            § 106 - Project approval and oversight
            § 107 - Acquisition of rights-of-way - Interstate System
            § 108 - Advance acquisition of real property
            § 109 - Standards
            § 112 - Letting of contracts
            § 114 - Construction
            § 118 - Availability of funds
        Chapter 3 - Other Provisions
            § 313 - Buy America

    Title 40: Public Buildings, Property, and Works
        Subtitle I, Chapter 11 - Selection of Architects and Engineers
            § 1101-1104 (Brooks Act): Selection of Architects and Engineers
        Subtitle II, Chapter 31, Subchapter IV - Wage Rate Requirements
            § 3141-3148 (Davis-Bacon Act)

    Title 41: Public Contracts
        Subtitle I - Federal Procurement Policy
            Chapter 33 - Planning and Solicitation (§§ 3301 - 3312)
            Chapter 37 - Awarding of Contracts (§§ 3701 - 3708)

    Title 49: Transportation

Code of Federal Regulations
    Title 2: Grants and Agreements
        Chapter II - Office of Management and Budget Guidance
            Part 200 - Uniform Administrative Requirements, Cost Principles,
                and Audit Requirements for Federal Awards (§§ 200 - 200.521)

    Title 23: Highways
        Chapter I - Federal Highway Administration, Department of Transportation
        Subchapter B - Payment Procedures
            Part 172 - Procurement, Management, and Administration of
                Engineering and Design Related Services (§§ 172.1 - 172.11)
        Subchapter G - Engineering and Traffic Operations
            Part 620 - Engineering (§§ 620.101 - 620.203)
            Part 625 - Design Standards for Highways (§§ 625.1 - 625.4)
            Part 626 - Pavement Policy (§§ 626.1 - 626.3)
            Part 627 - Value Engineering (§§ 627.1 - 627.9)
            Part 630 - Preconstruction Procedures (§§ 630.102 - 630.1110)
            Part 633 - Required Contract Provisions (§§ 633.101 - 633.211)
            Part 635 - Construction and Maintenance (§§ 635.101 - 635.507)
            Part 636 - Design-Build Contracting (§§ 636.101 - 636.514)
            Part 637 - Construction Inspection and Approval (§§ 637.201 - 637.209)
        Subchapter H - Right-of-Way and Environment (§§ 710 - 777)

    Title 29: Labor
        Subtitle A - Office of the Secretary of Labor
            Part 1 - Procedures for Predeterminiation of Wage Rates (§§ 1.1 - 1.9)
            Part 3 - Contractors and Subcontractors on Public Building or Public Work Financed in
                Whole of in Part by Loans or Grants from the United States (§§ 3.1 - 3.11)
            Part 5 - Labor Standards Provisions Provisions Applicable to Contracts Covering
                Federally Financed and Assisted Construction
                Subpart A - Davis-Bacon and Related Acts Provisions and Procedures (§§ 5.1 - 5.17)
                Subpart B - Interpretation of the Fringe Benefits Provisions of the Davis-Bacon Act (§§ 5.20 - 5.32)
        Subtitle B - Regulations Relating to Labor

    Title 40: Protection of the Environment
        Chapter V - Council on Environmental Quality (NEPA) (§§ 1500 - 1599)
            Part 1501 - NEPA and Agency Planning (§§ 1501.1 - 1501.8)
            Part 1505 - NEPA and Agency Decisionmaking (§§ 1505.1 - 1505.3)
            Part 1506 - Other Requirements of NEPA (§§ 1506.1 - 1506.12)

    Title 48: Federal Acquisition Regulations
        Chapter 1 - Federal Acquisition Regulation (FAR)
        Chapter 12 - DOT Transportation Acquisition Regulation (TAR)

    Title 49: Transportation

FIGURE 7.1   Sources of federal regulation, abbreviated mapping.

with cost principles contained in the Federal Acquisition Regulations of part 31 of title 48, Code of Federal Regulations.

Again, the purpose of providing these details is to inspire appreciation for the regulatory challenges of public sector development. Collectively, federal regulations dictate how public sector developers, particularly for federal projects or projects using federal funding, must: define terms; determine need, including design requirements; pursue funding and establish timing requirements; solicit and evaluate bids; negotiate and award contracts; specify contract clauses pertaining to sources of labor, subcontractors, and materials; manage contract performance; and guide many other aspects of the development process. Similarly, public sector projects at the state level must comply with state regulations, which are often equally complex and prescriptive. In instances where a state project benefits from federal funding, such as a NHS road project, the state must comply with both relevant state and federal regulations. This serves to add another layer of complexity to any project.

7.1.2. Predetermined Sites

Federal, state, and local governments are large landowners. The federal government owns almost 650 million acres of land, which is equal to approximately 30 percent of the total land area of the United States.[4] Though much of this is dedicated for conservation, such as the 84 million acres of parkland supervised by the National Park Service, a significant portion is not. States also own land, originally granted in most cases by the federal government at the time the state was first formed. In turn, states have passed control of some land to local governments and municipalities.

The relevance of the government's land portfolio is tied to its decision making as a developer. While private sector developers must identify and compete for sites in a market environment, public sector developers do not always need to do so. Whenever possible, public sector developers will use existing public land or existing government-owned buildings for a new project. This reduces project costs by eliminating the need to purchase land or easements. However, there may also be negative side effects if the predetermined project site is not ideally located.

For some projects, regardless of whether or not the public sector already owns land, a specific subject site is predetermined by necessity. This tends to apply more often to infrastructure projects than to building projects. For example, if the government needs to widen a road, land adjacent to the existing road must be acquired in order for the project to take place; no other existing, government owned land can be substituted because of the strict geographical requirement. Similarly, if the government is building a new connection route, there may be limited flexibility in determining the path of the new road. As described in Chap. 2, federal, state, and local governments have the right to expropriate private real estate for public use through condemnation. For example, 40 USC § 3113 permits acquisition by condemnation for the federal government while state powers of condemnation stem from state legislation.

Acquiring the necessary property, even when it is not being marketed for sale, is not an obstacle for the government from a legal perspective; however, it can certainly be problematic from both a political and financial viewpoint. Public sector takings can hardly be described as welcome and citizen outcry can turn an already complicated infrastructure project into a political battle between the agency responsible for the development and the elected officials supporting their constituents. Even when met with little opposition, use of eminent domain for condemnation can create financial pressures for public sector developments. The government is required to pay a fair price for any compulsory takings. If the land needed for a project happens to be valuable, the cost of condemnation may be substantial. Unlike a private sector developer, the public sector developer cannot simply decide the project is cost prohibitive and decide to look for another site.

7.1.3. Funding Considerations and Public-Private Partnerships

Public sector developers do not use traditional loans to finance their projects and, thus, do not have to seek debt financing from lenders as their private sector counterparts must. However, financing public sector projects comes with a unique set of problems. At the federal level, project funding generally comes from appropriations of tax revenue. Funding for public sector development projects at the state and local levels comes from one of three sources, depending on the type of project: (1) revenue from the issuance of bonds, (2) appropriations from the state's tax revenue, or (3) grants of federal funds, such as from the Highway Trust Fund described in Chap. 2. Note that some federal programs function as infrastructure "banks" and use federal funding to assist states and/or municipalities with specific types of public sector projects, such as certain water and wastewater projects. These programs/entities can loan money to states under extremely favorable terms from a revolving fund that is replenished when the borrowed funds are repaid (and then used to fund subsequent projects).

Regardless of the specific form, obtaining public sector funding for development projects involves a lengthy approval process as part of "acquisition planning" that requires the public sector developer to submit project budget estimates well in advance of undertaking the project itself—often years. In many cases, agencies are asked for multiyear plans anticipating future projects and the associated funding needs. While it is understandable that the public sector must engage in long-term budgetary planning exercises, accurate estimates for the cost of materials and labor for future development are difficult to predict so many years in advance. The obligation to engage exclusively with projects on an all-cash basis without a traditional third-party lender also means the government (at any level) can only fund a limited number of projects in a given year. Further, in this context, approval for development projects and must compete for funds with other spending priorities. For these and other reasons, federal, state, and local public sector developers can all find themselves facing budget shortfalls when the time comes to actually begin a development project. Inaccurate predictions are not the only sources of financial pressure for public sector development projects. As highlighted in Chap. 2 with respect to the American Society of Civil Engineer's (ASCE) Infrastructure Report Card and state of funds availability, such as with the Highway Trust Fund, the United States' infrastructure needs are quickly outstripping available sources of funding.

Privatization and public-private partnerships are both methods used with increasing frequency by the public sector to offset shortfalls in development funding needs. *Privatization*, as the name suggests, refers to the complete outsourcing of a development to the private sector. In instances of privatization, both the ownership and management of the resulting asset are held by the private sector. For example, rather than develop its own office building, a public agency may decide to become the rent-paying tenant of a private sector developer, who will develop, own, and operate/maintain a build-to-suit office project for the agency. While the government will still have to appropriate funds equal to the value of the lease, it would not have to budget for and undertake maintenance, repairs, and other expenses. In the context of infrastructure projects, privatization means allowing a private sector developer to build and operate a utility, which includes the collection of fees as a revenue stream. In this case, the public sector discharges its obligation to provide services to citizens by outsourcing to the private sector.

Public-private partnerships, also called PPPs or P3s, are projects in which a public sector government agency works in partnership with a private sector developer in order to complete a public-sector project. There are several different kinds of public-private partnerships, the structure of which can be fairly simple or extremely complex. Figure 7.2 describes several types of P3 structures. P3s are a procurement option, not a source of revenue or funding for the public sector. Note that in a P3, the private sector does not pay for the project, but merely finances it. P3s reallocate risk and responsibilities between the public sector and private sector, often leading to cost and time efficiencies, as well as technical innovations not previously accessible to the public sector. Importantly, P3s differ from privatization as the public sector retains ownership of the asset. Additionally, once a P3 contract is completed the asset is returned to public sector control. The only exception to this is detailed in Fig. 7.2.

In all P3 arrangements, both the private sector developer and the public sector agency make important contributions to the project. Typical public sector contributions include government

Traditional Procurement: Design-Bid-Build (DBB) contracts are considered the conventional procurement process where design and construction often are awarded to separate firms.

Design-Build (DB): Design and construction activities are bundled together; therefore the risk of coordinating the activities is transferred to a single concessionaire and can be considered a type of P3.

Design-Build-Finance-Operate-Maintain (DBFOM): Project responsibility and risk for design, financing, construction, operations and maintenance are placed on the private sector for the life of the P3 contract. DBFOMs are commonly found in projects with clear revenue streams, using projected revenue to attract equity and to leverage debt financing.

Design-Build-Finance-Maintain (DBFM): DBFMs are very similar to DBFOMs, with the exception that operations are held within the public sector. This contract is commonly used for social infrastructure projects such as hospitals, higher education infrastructure, and other buildings.

Design-Build-Operate-Maintain (DBOM): DBOMs bundle design and construction together with operations and maintenance of the completed asset. DBOMs are common in transit projects where the life-cycle costs are high due to maintenance demands of the asset. These projects can also take the form of a **Build-Operate-Transfer (BOT)**, where the private sector provides no financing, but constructs the asset, and upon construction completion transfers the asset to the public sector. This is often followed by a separate **Operate-Maintain (O&M)** contract.

Long-term Lease Concession: Commonly used for toll roads, this type of P3 is used to defease publicly held debt on the facility through fees paid by the private concessionaire. The private concessionaire is expected to provide maintenance, and possibly capital repairs to meet safety expectations and condition issues. **Operate-Maintain (O&M)** contracts also are common for existing assets, to provide private sector expertise to maintain an asset. Payments are made on through a fixed fee or incentive based model.

Build-Own-Operate (BOO): BOOs are the one exception to public ownership in the P3 model. BOO contracts are done at the encouragement of the government, through financial incentives, but the private partner retains ownership and operations of the facility.

FIGURE 7.2   Sample P3 structures.

owned land, concessions, permissions, or tax incentives. Revenue collected by the private sector developer comes most commonly from user fees, less so through availability payments, and rarely from shadow toll payments. Under an availability payment scheme, amounts are paid regularly through either prearranged or performance based payments. Shadow tolls, known in the United States as "pass-through tolls," are used more extensively internationally. Shadow toll payments are made to the private partner based on a per user basis. Common P3 contracts are variably structured with the potential for the private sector developer to provide financing (from equity investors and lenders), planning and construction services, and management or operation of the asset. Thus, a P3 can increase the capacity and speed of access to financing and access to technical expertise, resulting in a project with an accelerated delivery time line while reducing public cost and risk. Note that not all states have enacted P3 enabling legislation.

7.1.4. Political Will and Changes in Elected Office Holders

The government's role in serving its citizens is directed by elected officials at the federal, state, and local levels. In the context of development, *political will* refers to the degree to which these officials, individually and collectively, support a project as a matter of priority and are willing to commit resources to it. This often means allocating funds, but may also involve advocating for needed approvals, supporting related policy changes, and championing the project despite citizen opposition or party dissent. Such support is not necessarily tied to what any individual government agency considers to be a priority. The willingness of elected officials to support projects depends in part on the nature of the project itself but can also be heavily dependent on a politician's individual beliefs or priorities, lobbying influences, current citizen responses, reelection considerations, and other competing priorities. Perhaps understandably, political will can seem both enigmatic and fleeting.

Arguably, the two main reasons for changes in political are elections and reprioritizations due to external events. As previously established, public sector development planning is a lengthy process that can span several years, particularly for large, expensive, and complex projects. Such a protracted timeline inevitably spans at least one, if not more, election cycles. In these cycles, election-oriented behavior may cause a previous supporter to withdraw his/her support in order to focus on more visible or "relevant" issues intended to boost reelection chances. Championing government spending to build capacity at wastewater treatment plants, though perhaps critically necessary to community stability and growth, is unlikely to win a candidate the election if their opponent focuses on high-profile issues like safety in schools, job creation, or tax relief. If a project supporter loses an election bid, it can equally condemn a project. For example, prior to an election, there may have been widespread support for the development of a new bridge, money allocated, and the responsible agency may have begun soliciting bids for construction. However, if the election results in the change of office for at least one key political supporter and the newly elected official thinks the project is unnecessary and represents a poor use of funds, he/she may direct the agency to not award the construction contract or use other influence to end the project. In such a scenario, all the previous momentum and political will have been lost.

Elections are not the only thing that dampens political will. Unforeseen events can also disrupt prior support by creating the urgent need for the government to reprioritize its use of funds and other resources. If a severe storm damages roads and utilities, these obviously need to be repaired as a matter of priority and any planned new projects may not be pursued until after the storm damage has been addressed. Of course, not all causes of reprioritization are as clear. Opaque or seemingly unfounded policy reprioritizations can be a source of frustration for public agencies trying to advance their projects in order to provide services to citizens. This is especially true if a project is terminated after some, or perhaps significant, funds have already been expended for project preparation or design. Such funds are not recoverable and contribute to discussions about political waste.

7.1.5. Requests for Proposals and the Bidding Process

All levels of government are concerned with ensuring that interested and qualified candidates have a fair and equal opportunity to bid on government contracts. This is referred to as "full and open competition." Mandated procurement practices, specifically those involving soliciting offers and awarding public sector contracts, are the vehicle by which the government achieves full and open competition. At the federal level, this is prescribed across different parts of the USC and FAR; at the state level, procurement law and competition practices are embedded in each individual state code. Both state and federal statutes include requirements that certain opportunities are made available for small-, veteran-, and minority-owned businesses.

As part of the procurement planning process, the government agency intending to award a contract will assign a certified specialist who will be responsible for the solicitation, award, and administration of the contract. At the federal level, these specialists are called contracting officers (CO), but they may carry other titles at the state level such as procurement officer, bid officer, or even project manager. Only COs (or their state equivalents) are able to enter into legally binding procurement contracts on behalf of the government.[5] In the interest of fair competition, COs are required to publicly advertise government contracting opportunities. This obligation is met by posting the appropriate announcement, usually a request for proposals (RFP), on state or federal procurement websites. Federal opportunities will always be posted on FedBizOps.com; most states operate their own procurement websites. In order to bid on public sector contracts, firms must be registered in the appropriate system(s).

Although all levels of government are able to award a "sole source" contract to a specific firm in which no solicitation or bidding takes place, this requires justification and is only allowed under certain circumstances, such as to fill an urgent public need. The majority of public sector contracts must go through the solicitation process. The government's two primary solicitation methods are the use of sealed bids and a competitive bid process. In either instance, nonconforming bids are not considered. Sealed bids are most appropriate for procurements in which the requirements are clear and there is little need for discussion between the parties. Sealed bid contracts are awarded to the lowest price bidder that is competent to perform the work. Competitive bids are used for more complex public sector projects and are usually awarded based on the government's determination of either the "lowest price technically acceptable" or "best value."

Best value source selection allows the public sector to recognize the need to account for past performance when contracting for highly complex projects, particularly construction projects.

Federal construction and design (architect-engineering) contracts are governed by FAR Part 36, which allows for the use of two-phase design-build selection procedures and other specialized provisions.[6] It also allows architect-engineering vendors to be evaluated on the submission of technical qualifications as part of the government's process of developing a list of final bidders.[7]

The public sector's bidding process can be lengthy and complicated, requiring potential contractors to develop competency with bidding systems, government submission formats, and other procedures. Further, the government can modify its solicitations any time prior to bid submission, is under no obligation to award a contract or issue a start work order even after award, and has the unilateral right to terminate contracts for convenience at any point during the performance period. All these factors create risk of loss for private sector firms and make government contracts unappealing to those firms unwilling to commit the necessary resources required to work on public sector development projects.

7.2. TYPES OF PROJECTS

The public sector supports its citizens by providing and administering a wide range of services. This requires an extensive physical platform of both infrastructure and building projects, as outlined in the description of the built environment in Chap. 2. Public sector developments may serve direct citizen use, such as roads; facilities necessary for the provision of services, such as waterworks and distribution systems; or buildings to accommodate the government workforce, such as agency offices. Regardless of kind, each different type of use represents an area of public sector development.

7.2.1. Infrastructure

Federal, state, and local municipal authorities are responsible for the development and operation of infrastructure projects, without which society would struggle to function. These can include the following:

- Waterworks and distribution systems
- Sewage treatment works
- Dumps and solid waste (trash) facilities; recycling facilities
- Power generating plants
- Roads of all types
- Bridges
- Dams
- Runways and airfields
- Harbors and ports
- Parking structures for cars, buses, and specialty vehicles

Although many of these assets are more numerous at the state and local levels, the federal government does have its own portfolio of such infrastructure, often in a military context.

7.2.2. Buildings

All levels of government require building facilities to support public sector employees, services, and initiatives. These can include:

- Agency office buildings
- Barracks and other housing
- Medical facilities and hospitals
- Laboratories and test facilities
- Schools and training facilities
- Museums
- Data centers
- Airports and hangers
- Prisons
- Warehouses and storage facilities
- Industrial facilities

Additionally, state and local authorities also need buildings for things such as:

- Police stations

- Fire stations
- K-12 schools
- Community centers

To offer an example of scope, the federal government reported owning 498.7 million square feet of office space in 2016.[8] This single-user portfolio is greater than the entire office space inventory of the city of Chicago, which totaled 243.7 million square feet for the same year, including both the downtown and suburban markets combined.[9] The federally-owned office portfolio is largely equivalent to the office space inventory of New York City, which was 449 million square feet across the central business districts and 550 million square feet overall in the same year.[10] The government's reported total owned portfolio across all types of buildings, excluding land and infrastructure, was 2.399 billion square feet for 2017.[11] Note these figures represent only owned building space and do not include the federal government's lease portfolio. Nor do they include any state or local portfolios, whether owned or leased.

As with infrastructure, the creation, rehabilitation, and/or replacement of these structures each represents a potential public sector development project.

7.3. DETAILED CASE STUDIES

The scope and nature of public sector projects are often unique, which can create challenges for the private sector contractors involved in delivering these projects. The following sections will outline two public sector development case studies to explore some of the relevant issues in greater detail. The selected projects present an infrastructure project and a building project: the Woodrow Wilson Bridge and the Fort Belvoir Community Hospital, respectively. The case studies will provide an overview and timeline of each project as well as a description of unique public sector considerations and challenges.

7.3.1. Fort Belvoir Hospital

Fort Belvoir is a U.S. Army garrison, or "installation," located in northern Virginia. It was originally transferred by Congress in 1912 to the then War Department and has been occupied by the

U.S. Army since 1915 when it served as an engineer training facility.[12] The roughly 8,600-acre property is now owned by the Department of Defense (DoD) and is part of the Northeast Region of the Army's Installation Management Command (IMCOM).[13] Among other facilities, the garrison includes offices, housing, educational facilities, a commissary, and a hospital. More than one-third of the installation's total acres are preserved as designated wildlife sanctuaries.[14]

Fort Belvoir's current Community Hospital, which opened on the installation in 2010, serves all three branches of the U.S. Military. It is a 900,000 square foot (sf) LEED Gold modern facility, the construction of which was mandated as part of the federal government's Base Realignment and Closure (BRAC) process. The time between the design contract award and the occupancy of the hospital was only four years. Further, the project was undertaken in the context of the greater BRAC effort, which had considerable impact on surrounding parts of northern Virginia. The project's quick delivery, use of evidence-based design, and compliance within the overarching federal regulatory framework represent the type of superior results that can be achieved under an aligned team with motivated leadership and a clear vision. Nonetheless, this achievement was not without challenges. The following sections of this case study will examine the Fort Belvoir Community Hospital project in greater detail.

Background.   On May 13, 2005, a recommendation of the then current cycle of the military BRAC exercise was made to close the existing DeWitt Army Community Hospital (located on Fort Belvoir) and the Walter Reed Army Medical Center in Washington, D.C., and combine the two facilities into a new and expanded Community Hospital of more than 900,000 sf, to be located on Fort Belvoir. The 2005 BRAC mandate also included an order for the construction of several other new projects in and around Fort Belvoir, including, among others, the following (given sizes are rounded):

National Geospatial-Intelligence Agency campus: 2,400,000 sf

Dental clinic: 23,000 sf

Northern Regional Medical Command (NRMC) offices: 50,000 sf

National Museum of the United States Army: 300,000 sf

Wounded Warrior recovery center: 248,000 sf

Washington Headquarters Service (WHS) offices: 1.8 million sf at MARC center

The associated growth was anticipated to increase base personnel by as many as 22,000 people, raising the total from approximately 23,000 to 55,000.[15] Collectively, these BRAC projects created a tremendous amount of turbulence and had a massive overall impact on the surrounding communities; job location for military personnel and government contractors; and existing transportation infrastructure, both on-base and along Virginia roads. Adding to the turmoil was the fact that the BRAC recommendation, which became a legal mandate in November of 2005, included a deadline for the completion of the projects of September 15, 2011.

The Fort Belvoir Community Hospital project itself was governed by "BRAC 169." The entire development occurred within a Department of Defense owned property; however, water and sanitary sewer capacity was provided by Fairfax County. This had implications for the approval authorities for different elements of the project. Project planning and design exercises, which concluded in 2007, determined that the hospital would be built on an area that was occupied by the garrison's South Post nine-hole executive golf course. No significant wetlands or environmental concerns were associated with the site, although substantial existing infrastructure had to be relocated because of the preferred orientation of the hospital.

Two joint ventures (JV) formed a critical portion of the project team, with (1) a JV of firms HDR and Dewberry providing architecture and engineer services and (2) a JV of firms Turner Construction and Gilbane Building Company acting as construction managers (CM). Within the design/engineering JV team, Dewberry was responsible for site work and the design of an on-site central utility plant and parking structures, while HDR provided design and engineering services for the hospital structures. However, in the interest of time, the two firms worked collaboratively on most aspects of design and there was extensive integration and coordination of activities. The primary DoD authorities involved with the project were U.S. Army project leadership at Fort Belvoir garrison, the U.S. Army Corps of Engineers (USACE) and the Army Health Facilities Planning Agency (AHFPA), although the military's Tricare Management Activity, Portfolio Planning & Management Division (TMA-PPMD), and Medical Service Management Office (MSMO) also played important roles in the project. Several other agencies, both federal and nonfederal, had various project review authority.

The project featured the use of evidence-based design, which is "the application of building design elements known to facilitate physical, mental, social well-being, and productive behavior in its occupants and improved financial performance for its owners."[16] In the Community Hospital project, this led to innovative design changes such as different connectivity between spaces, decentralized care stations, changes to room layouts and fixtures, single-occupancy rooms that include guest space, and the incorporation of new technology and equipment.

Project Timeline. The following section presents a timeline of both the Fort Belvoir Community Hospital project as well as some key infrastructure projects.[17] Although not part of the BRAC scope, improvements to transportation infrastructure surrounding Fort Belvoir were necessary to accommodate the massive increase in traffic to and from the base as a result of BRAC projects. In addition to nearly doubling the day-time population of the base itself, new projects such as the National Museum of the United States Army were anticipated to attract as many as 1 million visitors per year.[18] Indeed, the base required nearly $30 million in internal road and infrastructure improvements, including a new access point onto the garrison from

U.S. Route 1.[19] However, funds were not provided for improvements on the surrounding road networks. This left state and local leaders searching for a solution to a serious road capacity problem created by a federal government decision (BRAC) that was made regardless of local impacts.

May 13, 2005:	BRAC recommendations are proposed with substantial impact on Fort Belvoir.
August 15, 2005:	Fort Belvoir's internal BRAC Implementation Plan is delivered to garrison leadership.
August/ September 2005:	Throughout the months of August and September, military leadership from Fort Belvoir engaged in various calls, Town Halls, briefings, and other outreach efforts with Virginia state and local leaders; the transportation impact of BRAC was consistently raised as an area of concern.
September 29, 2005:	The revised and final BRAC Implementation Plan is submitted.
November 9, 2005:	The BRAC recommendations become legal mandates; the DoD must begin closing and realigning affected properties within two years, by September 2007.
November 21, 2005:	USACE Baltimore District Office releases a request for proposal (RFP) for a BRAC Master Plan for Fort Belvoir.
February 17, 2006:	The USACE announces the selection of the Belvoir New Vision Planners (BNVP), a team of two primary firms and 17 subcontractors, including Dewberry, for the BRAC Master Planning contract.
March 31, 2006:	A five-year, $60 million contract is awarded to BNVP; a Strategic Communication Plan and Preliminary Siting Plan must be submitted within 60 days.

June 6, 2006:   HDR-Dewberry Joint Venture formed.

July 27, 2006:   The Army issues a press release describing the Preferred Siting Plan, based off of three siting plans contemplated by BNVP; the hospital will be built on the existing South Post Golf Course. Concerns about transportation infrastructure are immediately raised by the media and community.

August 2, 2006:   The BRAC projects' board of advisors meet; more than 14 transportation-related improvements are identified as necessary or critical. Military leadership commits to create a Transportation Working Group.

August 4, 2006: $2 million is included in the 2007 Defense Appropriations Bill for a traffic study to review the effect of BRAC in local transportation systems.

October 2, 2006:   Congress passes the 2007 National Defense Authorization Act with provisions for Fairfax County Parkway connector, $13 million for the construction of a Woodlawn Road connector.

January 8, 2007:   The Virginia General Assembly appropriates $12.5 million for grants to local jurisdictions impacted by BRAC. Of the total amount, $2.5 million was dedicated to traffic improvements around Fort Belvoir; $1.5 million was dedicated for "spot transportation improvements" around the base, and $1 million was designated specifically for the widening of Telegraph Road.

February 8, 2007:   Contract Task Order 1 awarded to HDR-Dewberry for the hospital.

February 13–15, 2007:   Four months of pre-design work culminate in the HDR-Dewberry Design Charrette; final site placement and design selected.

September 28, 2007:   USACE awards a $649 million contract to Turner-Gilbane for construction of the hospital; design is at approximately 10 percent.

December 18, 2007:   An agreement is reached between Virginia Department of Transportation (VDOT), the Army, and the U.S. Department of Transportation's Federal Highway Administration (FHWA) with respect to funding and constructing the Fairfax County Parkway connector, which provided access to and past Fort Belvoir. The four-part project was estimated to cost upward of $260 million, of which VDOT would be responsible for $114.7 million.

March 31, 2008:   The first concrete footing/foundation is poured for the new hospital.

October 31, 2008:   A groundbreaking ceremony was held to celebrate the construction of the Fairfax County Parkway connector improvements, although construction was not scheduled to begin until early 2009.

April 30, 2009:   The appropriation of $61 million in federal funding under the American Recovery & Reinvestment Act (ARRA) is announced to cover the cost of the remaining two phases of the Fairfax County Parkway extension.

December 11, 2009:   A "topping-out ceremony" is held to celebrate placement of the steel beam at the top of the hospital structure.

September 13, 2010:	A ribbon-cutting ceremony celebrates the opening of the expanded Fairfax County Parkway, which had to be widened to accommodate anticipated new traffic volumes. The road opened for traffic on September 19.
September 30, 2010:	Construction on Fort Belvoir Community Hospital is estimated at 66 percent complete with less than one year until the September 15, 2011 BRAC deadline.
July 27, 2010:	A ceremony is held at Walter Reed Army Medical Center to formally close the facility and transfer its functions to Fort Belvoir.
August 31, 2011:	Fort Belvoir Community Hospital opens; patients begin to arrive.
September 15, 2011:	BRAC completion deadline.
October 28, 2011:	The official ribbon cutting ceremony is held for the new hospital.
October 2011:	Construction on Belvoir and Pohick Roads completed; ongoing work to widen Gunston Road.
November 2011:	The DoD Office of Economic Adjustment provided funding of $180 million to widen U.S. Route 1 from Telegraph Road to Mount Vernon Memorial Highway.

Project Components and Costs. The Community Hospital project cost $1.03 billion and consisted of approximately 1.275 million square feet across several buildings:[20]

- Four outpatient clinics (500,000 sf)
- One central hospital building (700,000 sf)
- Two parking garages (for 1,458 cars and 1,759 cars) with 3,217 total spaces
- One central utilities plant (33,000 sf)
- One ambulance shelter (2,160 sf)

The final design included 120 state-of-the-art inpatient rooms, one intensive care unit, one cancer care center with two linear accelerators, surgical facilities including two da Vinci robots, and many other special treatment, imagining and therapy centers. The hospital, shown at completion in Figure 7.3, also included a helipad, ambulance motor pool facilities, and other standard elements of a modern hospital facility. Given the size of the project, a central utilities plant was needed on-site to provide heating, cooling, and emergency power.

Public Sector Considerations. As not only a public sector project but also a military project, the development of the Fort Belvoir Community Hospital had to be undertaken within the bounds of certain unique constraints. These included security considerations, appropriations limitations, political scrutiny, and specialized facilities development guidance. For example, construction took place on a secured military instillation, which had implications for workers' ability to access the site. The influence of Congress was another unusual consideration: the project had to be completed by the Congressionally mandated BRAC deadline and project budget was fixed by appropriation, which was subject to change. As part of its 2008 appropriations, Congress reduced the previously approved BRAC-wide budget by almost $3 billion, but subsequently also granted increases to specific projects in other instances, further complicating the budgeting process for an individual project.[21] A 2009 Government Accountability Office (GAO) inquiry into BRAC projects, including the Community Hospital and others at Fort Belvoir, is an example of the heighted scrutiny faced by public sector projects.[22] Finally, the planning, design, and construction of the project had to be in compliance with Unified Facilities Criteria (UFC) and the Unified Facilities Guide Specifications (UFGS), as per military directive.[23] The UFC and UFGS include guidance that is common for private sector projects, such as sustainable design, building systems and materials, and engineering requirements. However, they also require compliance with specialized military considerations such as Antiterrorism/ Force Protection (AT/FP) requirements as part of

12/02/11

FIGURE 7.3   Fort Belvoir Community Hospital.

the "commitment by DoD to seek effective ways to minimize the likelihood of mass casualties from terrorist attacks against DoD personnel in the buildings in which they work and live."[24] Such solemn considerations are rare for most private sector office and/or hospital projects. Some examples of specialized UFC requirements include:

- UFC 4-010-01 DoD Minimum Antiterrorism Standards for Buildings

- UFC 4-023-03 Design of Buildings to Resist Progressive Collapse

- UFC 4-510-1 Design: Military Medical Facilities

Project Challenges.   The following briefly outlines some of the more prominent challenges encountered during the Fort Belvoir Community Hospital project.

Challenge #1: Impact on Community.   At the time of the Community Hospital project (as today), Fort Belvoir was a major area employer of both military and civilian personnel. It was (and still is) surrounded by a developed community and bisected by several public roads that carry both through-traffic and serve the base's employment, delivery, and other transportation needs,

as shown in Fig. 7.4. Many existing roads were already estimated to be past capacity before any of the Fort Belvoir BRAC projects were undertaken.[25] Community members as well as state and local leaders shared concerns that the anticipated doubling of base personnel due to BRAC would overload road capacity, leading to delays onto and around the base. For example, the number of cars/people coming through the Fort's Pence Gate was expected to triple. There were also concerns about the implications for local school enrollment and capacity. From an economic development standpoint, BRAC activity increased demand by government contractors for 8 to 10 million square feet of new office in the surrounding area;[26] however, this growth represented additional potential strain on infrastructure systems. Under BRAC, neither the DoD nor other branches of the federal government were responsible for the cost of increasing infrastructure capacity in the surrounding areas. The cost burden fell, essentially without warning as a result of BRAC, to local and state authorities. Perhaps understandably, there was considerable negative press coverage surrounding the BRAC projects at Fort Belvoir.

To manage the situation, Fort Belvoir leadership hosted regular community outreach activities and met with local, state, and even national leaders to

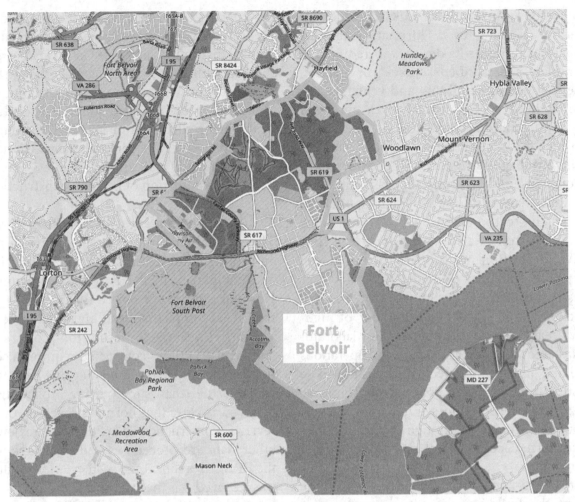

FIGURE 7.4   Fort Belvoir and surrounding road network. (Image created using data available under the Open Database License, © OpenStreetMap contributors.)

discuss BRAC-related projects and their impacts on the surrounding areas.[27] A working group was also convened to study the impact of BRAC on transportation issues with representatives participating from the U.S. Army, VDOT Northern Virginia District, Fairfax County, and the U.S. Department of Transportation FHWA Federal Lands Highway Division, as well as outside consultants.[28] Highway infrastructure was particularly problematic and triggered a BRAC-Related Area Plan Review by the Fairfax County Department of Planning and Zoning. Several critical roads were identified as targets for increased capacity and leaders at all levels sought to find funding for these related projects; the results of these efforts are reflected in the project timeline. New bus lines to serve Fort Belvoir were also added by the Washington Metropolitan Area Transit Authority.[29]

Infrastructure fundraising efforts and improvements continued well after the completion of the Community Hospital project and the BRAC deadline. Although these challenges were not connected to the project scope of work (SOW), they were materially relevant to the successful use of the project upon completion and were also relevant for the ongoing relationship between Fort Belvoir and the surrounding community. There can easily be a significant disconnect between high-level strategic public sector planning activates, like BRAC, and considerations for the impact on surrounding areas and/or systems if project visioning is done in isolation. Whenever possible, projects, especially large ones, need to be considered holistically within the context of surrounding communities, infrastructure systems, and other relevant factors.

Challenge #2: Security. The UFC classifies the Community Hospital as a "primary gathering building," which means its design was required to accommodate a prescribed setback (or standoff distance) between the building and vehicle access based on criteria such as threat level, perimeter security, and building construction. The hospital also had to be designed to resist explosions, which involved specialized design considerations and also the use of blast resistant glass and other threat-rated materials. Use of such specialized materials adds to the cost of a project. In some cases, these materials also have a long lead time for delivery, which, if not properly managed, can impact project timeline.

Access control affects design for people as well as vehicles. The hospital has a number of high-security areas, including the pharmacy, laboratories, data center, mechanical spaces, and operations center. Each of these high-security areas had to be designed with card key access control. The same is true for all exterior doors other than the primary public entrances, although even the public entrances have access control outside of normal operating hours.

Security was not only a consideration for the design of the building; the actual construction process itself was also affected by security-based concerns. As previously stated, construction of the Community Hospital took place on a secured military instillation. This had implications for basic operations, such as the need to screen construction crews and materials entering the site on a daily basis. Over 500 workers were at the construction site every day and, under normal operations, it would have taken hours to complete the required inspections.[30] This would obviously have reduced the amount of productive time for workers once they gained access to the construction site, but would also have led to an insupportable delay for normal base employees and visitors. To address this logistical dilemma, in January of 2009, a special solution was created in which base roads inside the Pence Gate were closed and special access points were created to allow crews and materials to access the site directly off the adjacent U.S. Route 1. This involved relocating a portion of the garrison's perimeter security such that the construction site was not secured and workers were, therefore, not subject to normal access control inspections at Pence Gate. Dewberry coordinated with Fort Belvoir security personnel and VDOT to construct a new, temporary intersection along U.S. Route 1. This required the construction of a right turn deceleration lane on U.S. Route 1 and the preparation of a comprehensive traffic management plan. This temporary access was closed upon the completion of hospital construction.

Challenge #3: Challenges under a BRAC Deadline. The normal timeframe for planning, designing, and constructing a military hospital of similar size and complexity to the Fort Belvoir Community Hospital can range from 10 to 12 years; however, the BRAC completion deadline for all projects was a firm five years. As a result, the project's design and construction contracts required completion by the BRAC deadline, but this had statutory weight rather than simply being a planning horizon. It was clear that normal processes for design reviews, construction shop drawing reviews, and other standard procedures had to change to meet the BRAC deadline. One such example of how this was achieved was the extensive use of partnering on the project. While many USACE projects involve partnering, few, if any, have had as extensive and productive a partnering effort as the Fort Belvoir Community Hospital. Partnering occurred at two levels on a monthly basis. The project leaders team included the leaders from the USACE, the Army Health Facilities Planning Agency, DeWitt Hospital, Design Project Managers, Construction Operations Managers, and key Fort Belvoir installation personnel. Issues that could not be resolved at the project leaders team were elevated to the project executive team, which had the executives of the same organizations including the USACE Norfolk district commander, director of HFPA, and principals from both the design firms and construction companies. Personnel at partnering meetings were empowered by their organizations to negotiate on process, risk, and procedure and reach agreements that were binding on the entire team. Partnering also led to real-time "over-the-shoulder" design reviews and sharing of Building Information Models (BIM), used to review, edit as necessary, and approve construction submittals. Without the committed participation of all parties involved in the partnering, the BRAC deadline could not have been met.

In addition to efficiencies achieved through partnering, a new delivery model was used in order to meet the BRAC deadline: Integrated Design-Bid-Build (IDBB). The commercial delivery method most similar to IDBB is the "construction manager at risk" format. In an IDBB system, both the designer and the GC are under contract to the owner but the construction contractor is brought on early for preconstruction services to comment on the design constructability, cost, and schedule. Construction can be phased, as it was on the Fort Belvoir Hospital project, so that work can begin as soon as possible. In the case of the hospital, design was at approximately 10 percent when work began. Site preparation and utility relocation started immediately after award of the construction contract and the construction packages went from the designer to the construction team in phases such as foundations, core and shell, and finishes. Long lead time materials were given special attention by the design team in order to finalize design and specifications with enough time to procure the necessary components, such as elevators, generators, electrical switch gear, chillers, cooling towers, and other major systems. This allowed the contractor to purchase these items and accommodate long delivery times even while the building itself was still being designed. Although the USACE had used IDBB on several smaller projects in the Kansas City District, nothing as large or complex as the Fort Belvoir Community Hospital had previously been undertaken. The process was also new to the industry. Four large construction companies attended all the informational "industry days" and pre-solicitation events held for the hospital project prior to bid. Each of these four large companies had sufficient bonding capacity; however, because IDBB was a new contract process for this size project, bonding companies strongly favored construction joint ventures to spread risk.

Summary/Conclusions. The $1.03 billion Fort Belvoir Community Hospital was a high-profile development project faced with security challenges, inadvertent impact on the surrounding community,

and a tight, BRAC-imposed delivery deadline. Further, the project also had to meet unique government requirements, such as compliance with the Unified Facilities Criteria. The use of an innovative IDBB project delivery method and active communication allowed the development team to overcome these challenges and successfully complete the Fort Belvoir Community Hospital within the allotted BRAC time frame.

7.3.2. Woodrow Wilson Bridge Case Study

The Woodrow Wilson Memorial Bridge is a drawbridge spanning the Potomac River connecting Virginia and Maryland. The bridge is a part of Interstates 95 and 495, making it part of the NHS. It originally opened in 1961 and was later replaced by a newer, larger bridge (of the same name) that officially opened in 2006 and was completed by 2013. The original Woodrow Wilson Bridge structure was constructed in four years for a total cost of $14 million. In contrast, the planning and construction process for the current (replacement) Woodrow Wilson Bridge project took 20 years and cost more than $2.5 billion, the cost of which included not only the bridge but also the approaches and interchanges necessary to integrate it with the existing road system. Even adjusted for inflation, this increase in cost (and time) is significant.

The two-decades long replacement process is testimony to the challenge federal, state, and local regulations and local politics present to replacing major infrastructure in the United States today. The project experience also speaks to the fundamental importance of adopting an inclusive approach to, and investment in, project coordination, legal assistance, environmental mitigation, and public outreach. The following sections of this case study will discuss the timeline and challenges of the Woodrow Wilson Bridge replacement project in greater detail.

Background. When the original, federally owned Woodrow Wilson Bridge opened in 1961, it had six lanes and a design capacity to support 75,000 car crossings annually. An estimated 14,000 vehicles per day crossed it during its first year. The bridge was designed and constructed as the inner circumferential loop around Washington, D.C., and was intended to be part of a proposed comprehensive regional network that included two additional loops as well as the extension of I-95 through the District of Columbia. Under this scheme, the bridge would be part of the I-495 beltway "inner-loop" but be distinct from I-95, thus allowing more flexibility and greater traffic flow around the district.

However, the District or Columbia subsequently vetoed construction of its portion of I-95 and other suburban jurisdictions followed suite, vetoing construction of the two other planned loops. Thus, the Woodrow Wilson Bridge was left as the only Potomac River crossing and inherently became part of both I-495 and I-95. It was (and still is) the only route for not only local and regional traffic, but for all east coast north-south through traffic crossing the Potomac River in the southern part of the Maryland-Virginia-District of Columbia "National Capital Region." Even today, the nearest downstream Potomac River crossing is the U.S. 301/Governor Harry Nice Bridge some 40 miles to the south.

Only a few decades after its opening, traffic approached 200,000 cars per day on the Woodrow Wilson Bridge, far exceeding the anticipated volume. Accidents increased, safety declined, and pavement and structural deterioration occurred much earlier than predicted. Further complicating matters were the inherent flaws in the legacy design of the bridge: as the only one of the three loops originally planned in the road network, the six-lane Woodrow Wilson Bridge was forced to absorb and merge traffic from the widened eight-lane Capital Beltway, causing congestion. Regional and interstate congestion and delays where further exacerbated by the opening of the bridge's draw span upward of 200 times a year to accommodate the passage of ships destined for Alexandria, VA and the District of Columbia waterfront. By the mid-1980s, high traffic volumes, including heavy trucks, hastened structural deterioration and caused the FHWA to declare the bridge structurally deficient.

Project Timeline. The Woodrow Wilson Bridge replacement project spanned several decades.

A brief chronology of the planning and replacement construction process is as follows:

Mid-1980s

- The FHWA declares Wilson Bridge structurally deficient and functionally obsolete due to high daily traffic volume, structural condition, and inability to handle multimodal traffic uses.

1988

- FHWA initiates discussions/planning with Maryland, Virginia, and the District of Columbia.

1991

- The first Environmental Impact Statement (EIS) is published by FHWA with sponsorship from District of Columbia, Maryland, and Virginia. It studies five alternatives, each of which includes 12-lanes. Opposition from neighborhood groups and lack of local political support create the need for additional studies and community involvement in order to advance the project through the National Environmental Protection Act (NEPA) process.

1992–1995

- FHWA establishes a Project Coordinating Committee, which considers more than 350 suggestions and alternatives and recommends some for further consideration in EIS process.

- Congress enacts the Woodrow Wilson Memorial Bridge Authority Act of 1995 to facilitate bridge transfer to joint Maryland-Virginia-District of Columbia authority and provide initial funding federal funding for the project.

1996

- The Project Coordinating Committee endorses a "preferred alternative" featuring a two span drawbridge with 70-foot navigational clearance. Amended EISs are submitted in January and July.

1997

- September—a Final EIS is published that evaluates eight 12-lane alternatives including a high bridge and tunnel, both of which would have eliminated the need for a draw span.

- November—FHWA selects a preferred alternative consisting of two parallel drawbridges with six lanes, plus a bicycle/pedestrian facility.

1998

- A unique bridge design competition considers seven concepts from four firms and selects a final design in November. The winning project design team was overseen by Potomac Crossing Consultants (PCC) acting as the overall general engineering consultant.

- The City of Alexandria files a lawsuit contending that the project violates several National Environmental Protection Act (NEPA) provisions.

- Given unresolved litigation and other issues, the FHWA decides to prepare a supplemental EIS and creates an Interagency Coordination Group of 29 federal, state and local entities to expedite the resolution of numerous construction, environmental, historic preservation, and other EIS-related matters.

1999

- March—the City of Alexandria settles its suit with FHWA on the Final EIS, but litigation is pursued by community groups, including the Sierra Club, who have concerns for the environment and the disturbance of important cultural resources such as historic cemeteries.

- A district court subsequently rules in favor of the plaintiffs but is eventually overruled by the District of Columbia Circuit Court.

2000

- The Final EIS is completed and approved by FHWA.

- Potomac River dredging begins in October. Refined bridge designs result in the need for more extensive dredging and disposition of material than original estimates. Dredge material was used to restore the historic Shirley Plantation's agricultural base by filling a 50-acre site that had been mined for sand and gravel. Cost: $14.5 million

2001

- VDOT completes a $4.8 million steel bridge deck replacement of the bascule span to keep the deck operational until first new span is complete and opened for traffic.

- Bridge foundation work begins on more than 1,300 45-foot concrete columns. Cost: $125 million

- November—initial bids for the main bridge construction contract are received by Maryland State Highway Administration (MSHA) at $870 million, which was 70 percent over the original $400 to $500 million estimates. MSHA rejects this bid and initiates an Independent Review Committee (IRC) to review the bidding documents and plans. The IRC consists of national bridge design and construction experts.

- The District Department of Transportation, MSHA, and VDOT finalize designs and right-of-way acquisition and begin geotechnical testing programs for the reconstruction of interchanges at MD-210, I-295, and U.S. Route 1.

2002

- March—MSHA publishes the recommendations of the IRC to redesign and restructure construction bids for the main bridge to:

  o Increase competition by breaking out separate bid packages

  o Standardize unique design

◦ Address complicated erection issues

◦ Improve plan clarity

2003

- After a two-year delay the three restructured main bridge contracts are awarded, including:

  ◦ The Bascule Span Contract

  ◦ The Virginia Land Approach Spans

  ◦ The Maryland Water Approach Spans

2006

- The first span opens on May 18th. Two-way traffic is transferred from the old six-lane structure to the new six-lane span and the original bridge is demolished over the next year.

2008

- The second span opens on the weekend of May 30th to June 1st and the MD-210 and I-295 interchanges and the Virginia U.S. Route 1 interchange are open by year's end.

2013

- The Telegraph Road interchange in Virginia is completed. This is the final major element of the overall Woodrow Wilson Bridge project.

Project Components and Costs. The finished 12-lane Woodrow Wilson Bridge was officially dedicated in 2006 and opened fully to traffic in 2008, after two decades of struggle.[31] It consists of two parallel six-lane bascule spans with 70 feet of navigational clearance when closed. Each span provides for three lanes for local traffic access between Telegraph Road and MD-210, two lanes for express traffic, and a one lane reserved for a future HOV/bus/rail. The project also includes 7.5 miles of Capital Beltway (I-95/495) improvements the reconstruction of four interchanges. The bridge and surrounding road network are shown in Figure 7.5.

The total (approximate) project cost of $2.5 billion included $830 million for the bridge itself with the balance of project cost attributable to

bridge approaches, related interchanges, and environmental mitigation. The breakdown of major project costs (rounded) and associated contracts are shown below and contributions from the major jurisdictions (only) are highlighted in Fig. 7.6.

- Foundation contracts—$125 million

- Bridge crossing (3 contracts)—$830 million

- MD-210 Interchange (3 contracts)—$165 million

- I-295 Interchange (4 contracts)—$297 million

- U.S. Route 1 Interchange (6 contracts)—$623 million

- Telegraph Road Interchange (3 contracts)—$400 million

Public Sector Considerations. Funding was a major consideration for the $2.5 billion Woodrow Wilson Bridge project. Four different DOT agencies were involved in the project representing Maryland, Virginia, the District of Columbia, and the FHWA. These four agencies shared in the project costs, which necessitated the coordination of appropriations for project funds. Sufficient institutional knowledge existed between these agencies based on various overlapping infrastructure projects in the National Capital Region to facilitate the coordinate appropriations effort. The Woodrow Wilson Bridge specifically benefitted from a $900 million federal commitment of funds from the Highway Trust Fund through the Transportation Equity Act for the 21st century (TEA-21) and $600 million in funding through amendments to the original Woodrow Wilson Memorial Bridge Authority Act.[32]

Project Challenges. The following briefly outlines some of the more prominent challenges encountered during the Woodrow Wilson Bridge project.

Challenge #1: Project Coordination, Management, and Decision Making. The federal government owned the original Woodrow Wilson Bridge

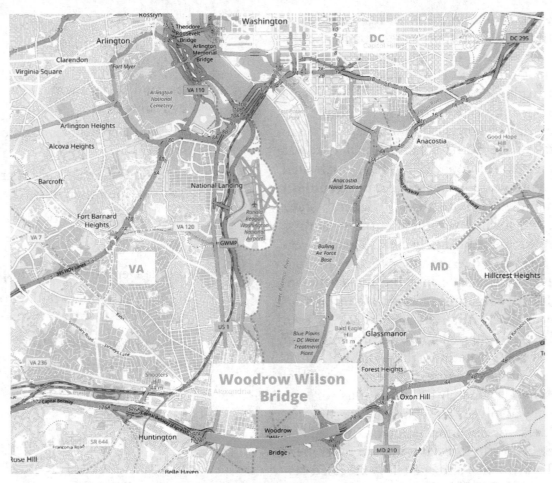

FIGURE 7.5   Woodrow Wilson Bridge Project area. (Image created using data available under the Open Database License, © OpenStreetMap contributors.)

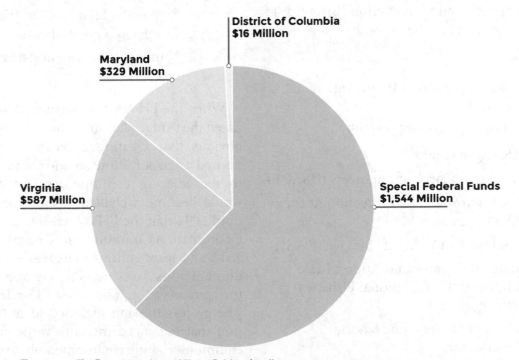

FIGURE 7.6   Woodrow Wilson Bridge funding sources.

structure yet, depending on their respective locations, different bridge approaches, spans, and the bascule span lie in Maryland, Virginia, and the District of Columbia. The district was responsible for bridge operations simply because the operable portion of the bridge lay within the district's boundaries; however, only a relatively small percentage of the traffic either originated or was destined for Washington, D.C. This complicated arrangement naturally created a complex project environment with more than 20 federal, state, regional, and local government entities involved in the replacement bridge project. This unprecedented level of government involvement does not necessarily translate into the most efficient method for coordinating a large-scale infrastructure project. Each different agency has its own teams of people, internal approval processes, and capabilities and limitations; these did not inherently align with the respective teams, approval processes, and capabilities/limitations of any other agency involved. The consequences of withholding, delay, or misinformation easily escalated and, during the first decade, many government bodies, departments, and agencies expressed displeasure at the overall lack of project communication and coordination.

To highlight the difficulty, the following is a list of different agencies and jurisdictions involved (at one point or another) in the process:

- Federal
  - Federal Highway Administration (FHWA)
  - National Park Service (NPS)
  - U.S. Coast Guard
  - U.S. Army Corps of Engineers (USACE)
  - U.S. Environmental Protection Agency (EPA)
  - U.S. Fish and Wild Life Service (FWS)
  - National Oceanic and Atmospheric Administration's National Fisheries Service
  - Advisory Council on Historic Preservation

- Maryland
  - Maryland State Highway Administration
  - Maryland Department of the Environment
  - Maryland Department of Natural Resources
  - Maryland National Capital Park and Planning Commission
  - Prince George's County, Maryland

- Virginia
  - Virginia Department of Transportation
  - Virginia Department of Environmental Quality
  - Virginia Institute of Marine Science
  - City of Alexandria, Virginia
  - Fairfax County, Virginia

- District of Columbia
  - Department of Public Works (DPW)
  - District of Columbia Health Department
  - District Department of Transportation

- Other
  - Regional—Metropolitan Washington Council of Governments
  - Numerous individual state and local officials

When the FHWA and partner jurisdictions initiated the bridge planning study in 1988, it became obvious that creating a balance regarding the needed transportation solution and the impacts to the adjacent communities and environment would become a challenge. However, it was not until 1992 that the FHWA created an Interagency Coordination Committee (or Group) of 29 federal, state, and local entities to expedite the resolution of numerous construction, environmental, historic preservation, and other EIS-related matters. The group's mission included identifying a solution that enhanced mobility while ensuring that community and environmental concerns were addressed. It also developed the following leadership teams to help coordinate decision making and implement the program:

- *Project Leadership Team:* Charged with providing strategic decision making, policy direction, and performance review. Consisted of the executive leadership of the District of Columbia, Maryland, and Virginia DOT's and FHWA.

- *Project Management Team:* Responsible for technical assistance and operational management. Consisted of the project managers of Maryland, Virginia, and FHWA and had charge of day-to-day operations of the project.

- *Environmental Management Group:* Responsible for completing Final EIS, securing permits, and monitoring environmental commitment compliance.

- *General Engineering Consulting Team:* A tri-venture of Parsons-Brinkerhoff (now WSP), URS (now AECOM), and RK&K.

- *Virginia Technical Coordination Team:* Reviewed design and project matters related to Virginia portion of project.

- *Interagency Coordination Committee:* As noted, involved more than 20 government agencies.

- *Design Review Working Group:* Sponsored the bridge design contest and reviewed other design matters as well as cultural and historic matters.

Challenge #2: Legal Issues and NEPA/Environmental Mitigation.

The 1969 National Environmental Protect Act (NEPA) and subsequent amendments impose substantial mitigation requirements on "major federal action(s)." The NEPA evaluation process, as outlined in the Code of Federal Regulations, is triggered when a federal agency develops a proposal for a significant project, such as the Woodrow Wilson Bridge. Of the three levels of NEPA review, the most stringent is the Environmental Impact Statement (EIS) review, which is required "if a proposed major federal action is determined to significantly affect the quality of the human environment."[33] The complexity of the NEPA regulation and its interpretation by the court systems frequently creates a dilemma with respect to determining what measures constitute compliance under different circumstances for different projects. Challenges to the adequacy of mitigation by environmental and community groups often result in costly project delays and even cancellation.

Through a 10-year planning and NEPA process, the Woodrow Wilson Bridge Coordination Committee recommended a "preferred alternative" in 1996 for the new bridge to be a (maximum) 12-lane facility consisting of side-by-side, movable span, twin bridges with a 70-foot navigational clearance. The improvements would include upgrades and reconfigurations of the four approach-interchanges such that they could accommodate both a local and express system as well as provisions for future HOV, express buss, and/or heavy rail. Further, a 14-foot pedestrian facility was specified. In 1997 the recommendation was reviewed by FHWA, which approved the EIS and issued a Record of Decision (ROD).

Specific concerns with respect to the Woodrow Wilson Bridge project included fears that the project would adversely harm the Potomac River, natural habitats, and wetlands both downstream and adjacent to the bridge given its new, larger footprint and extensive construction operations. There was also concern by stakeholders that a 12-lane bridge was unnecessarily large. These concerns generated national attention and a number of alternative reviews and recommendations, which included the following:

- Southern alignment—relocate the bridge to the south of the current alignment to decrease impacts on Jones Point Park and the City of Alexandria.

- Tunnel—construct a tunnel to eliminate the visual impacts of the bridge.

- Raise the bridge to remove the movable span.

Each of these alternatives were considered in isolation and in combination, but were dismissed due to environmental impacts, cost, and based on a standard of reasonableness to meet the purpose and need as stated in the EIS process.

Despite an unprecedented level of public and community involvement in the NEPA process, stakeholders remained concerned about the impacts that the selected alternative would leave on the adjacent communities. In 1999, the City of Alexandria and the Coalition for a Sensible Bridge filed suit in Federal District Court against the FHWA and the NEPA process that lead to the ROD. FHWA negotiated a series of mitigation packages with the City of Alexandria to address their concerns. The total $64 million package provided for recreational facilities, streetscapes near the project, and provided for a historic memorial to freed slaves whose remains had been buried adjacent to the project. A Federal Judge ruled in favor of the remaining plaintiffs in April of 1999, but was overruled in the U.S. Court of Appeals in December 1999; further appeals to the Supreme Court were denied, clearing the way for the project to proceed to final design and construction in 2000.

Ultimately, additional efforts that were incorporated into the project by the FHWA and project team, including the decision to create a comprehensive mitigation package for the Woodrow Wilson Bridge project that arguably went well beyond that which is typically undertaken. In addition to measures to protect water quality, fish, wildlife, and environmental factors immediately adjacent to the project, many off-site investments were made as well. These included stocking multitudes of fish in various tributaries, establishment of a bald eagle sanctuary, construction of a Chesapeake Bay fish reef, preservation or creation of nearly 150 acres of wetlands at various Maryland and Virginia locations, and planting of 20 acres of river grasses in the lower Potomac to improve fish habitats and water quality. Further, an independent party was hired to monitor progress and create a tracking system to ensure full compliance with the mitigation plan. These decisions had both cost and time ramifications for the project but proved instrumental in resolving legal actions in the late 1990s as well as in preventing future legal challenges.

Challenge #3: Coordinating the Opening of the New Woodrow Wilson Bridge. As early as the initial project planning stage, team members recognized that coordinating the opening of the new bridge and decommissioning the old bridge would be a challenge. Initial construction documents anticipated switching traffic to the new bridge in two distinct construction operations: (1) a mid-2006 switch from the existing bridge to the new six-lane "outer loop" spans and (2) demolishing the existing bridge and opening the new "inner loop" spans in late 2008. However, unique project conditions made this difficult to implement in practice, including:

- Different entities in charge of different elements of the bridge project based on their respective jurisdictions.

- Conditions that necessitated drastically different construction challenges on the different approaches.

- The volume of over 200,000 vehicles per day.

These issues were addressed through a multi-jurisdictional agreement between District of Columbia, Maryland, and Virginia stipulating that, upon completion, the project removed ownership of the bridge from the FHWA and gave joint operational responsibility to Maryland and Virginia on a 50/50 basis. This agreement also stated that if one government delayed the construction of the others' portion of the overall project, the delaying government would be responsible for all delay cost. This provided the incentive for Maryland and Virginia, the two governments with the most at stake in the timely completion, to work together in planning the traffic switch.

Challenges related to project timing were intensified due to the vastly different site conditions on the Maryland and Virginia sides of the bridge. The land needed for new bridge abutments and highway approaches on the Maryland side was predominately greenfield; conversely, in Virginia, construction had to be integrated into the existing I-95/I-495 infrastructure and the surrounding area was already heavily developed. To integrate the existing road system and support increased capacity for the new bridge, the U.S. Route 1 and Telegraph Road Interchanges in Virginia had to be reconfigured and reconstructed. The Telegraph Road Interchange alone, an award-winning project for which Dewberry served as the Section Design Consultant, required a widening and reconstruction of 2.5 miles of the Capital Beltway (I-95/I-495), 11 bridges/access ramps, 22 retaining walls, and 200,000 sf of noise barrier walls. The design and construction of this work had to be coordinated with the U.S. Route 1 Interchange work as well as with the bridge itself. To accommodate the Virginia-side improvements, a high-rise apartment building, two-story office building, and a number of garden apartments had to be demolished. This also required the relocation of over 260 Alexandria residents. Initially, this placed the Virginia timeline roughly 9- to 12-months behind the Maryland timeline with respect to being able to shift traffic to the new bridge.

However, VDOT took advantage of an innovative highway design change, created and refined by various members of the project team, to provide a temporary tie-end to the existing bridge from the widened highway. This allowed all work adjacent to the new bridge to occur outside of traffic. VDOT then introduced incentive contracting to ensure that all work on the Virginia approaches between U.S. Route 1 and the Virginia-side abutment was complete in order to meet the mid-2008 opening of the outer loop spans.

The actual traffic switches required regional coordination of interstate traffic from Richmond, VA to Baltimore, MD, in order to minimize delays and increase safety. To facilitate the work for the shifting of actual lanes (e.g., final paving, striping, and other roadside appurtenances installation), VDOT closed the Springfield Interchange heading north to all traffic and MSHA encouraged drivers to use other Potomac River crossings. Advance media efforts and traffic operations centers from Florida to Maine coordinated in an effort to decrease the number of vehicles utilizing I-95/I-495 to cross the Potomac River during the switch.

Challenge #4: Community Outreach, Education and Involvement. Concerted public outreach was absent during the first decade of the Woodrow Wilson Bridge replacement process, which may have contributed to the project's troubles in the 1990s. Although city-wide polling by the program manager revealed that the majority of citizens supported a new and enlarged bridge, opposition to the project came from environmental groups and local neighborhood associations. Tension between the "bigger picture/greater good" project perspective and the neighborhood perspective proved to be a serious challenge.

The level of community/political activism in the greater metropolitan Washington region is among the highest in the nation. In particular, the neighborhood associations of wealthy areas surrounding the bridge have active members comprised of attorneys, advertising and public relations professionals, Capitol Hill staffers, political campaign managers, and other well-connected individuals. Several of these associations contributed to negative media coverage of bridge proposals and political pressure to stall the project.

In the late 1990s, the project sponsors hired professional public relations experts to develop a comprehensive public outreach program that became critical to the project's success. The program included project information centers and open houses in Alexandria, VA and Prince George's County, MD. It also featured a project website, speaker's bureau, quarterly briefings for public officials, newsletters and factsheets, and updates on other activities to keep the public informed and involved. Such efforts were important not only to help bring public officials on board, but also to demonstrate to courts the project's commitment to go above and beyond in involving the public and considering and incorporating its input with respect to environmental concerns.

Private sector support and advocacy was also critical to the project's success. Organizations including the Greater Washington Board of Trade, American Automobile Association, and Northern Virginia Transportation Alliance provided critical assistance in building a strong public record of support based on the EIS's findings and crucial

FIGURE 7.7   Woodrow Wilson Bridge. (*Courtesy of Virginia Department of Transportation* [VDOT].)

encouragement to elected officials to support the project based on its well-documented merits.

Summary/Conclusions. The Woodrow Wilson Bridge replacement project succeeded, after a decade of pursuing a more traditional process, by recognizing the need to adapt to existing regulatory and political pressures. Given the many engineering consultants, construction contractors, and agencies involved, special efforts were made to create a project coordination group and assemble experts and to improve commitments to environmental mitigation, innovative contracting, and community outreach. Through their combined endeavors, the project team was successful in building consensus around an award-winning design, securing EIS approval, and ultimately constructing a replacement bridge (along with significant highway and interchange improvements) that has become a regional and national landmark. Figure 7.7 shows the Woodrow Wilson Bridge today. However, if investment in greater community outreach, education, and involvement had been made earlier in the process, the bridge replacement may well have been constructed sooner and at a lesser overall cost.

CHAPTER CONCLUSION

Public sector development projects are subject to most of the same challenges as private sector projects, but they also face unique circumstances. Project inception for the public sector is often tied to an evaluation of public needs, appropriations, and compliance with an extensive regulatory environment with respect to pre-bid and contracting requirements. Both private and public sector projects face the same site concerns with respect to location and access. They also share technical due diligence concerns, including the impact of environmental, topographical, and geotechnical issues on project design, cost, and timeline. However, the public sector is often subject to additional reviews, such as those imposed by NEPA, particularly if a state or local entity uses federal funding. Public sector developers are not necessarily concerned with market demand for their products in the same way as private sector developers are; however, they are affected by market pricing if compensation must be paid to condemn sites or acquire right-of-way easements. The public sector has increasingly turned to the private sector, through the use of public-private partnerships, to complete public sector projects more cost effectively and efficiently. Combining the strengths of both public and private developers to create projects enhances the built environment in ways that would not have been possible otherwise.

REFERENCES

1. FAR 2.101: Definitions.
2. State of Ohio Procurement Handbook for Supplies and Services, Chapter 11: Construction, Repairs, Maintenance; Ohio Department of Administrative Services, General Services Division, Office of State Procurement, available online at https://procure.ohio.gov/pdf/PUR_ProcManual.pdf (accessed December 9, 2018).
3. Ohio Department of Transportation website: http://www.dot.state.oh.us/Pages/Home.aspx
4. U.S. Geological Survey, The National Map website, available at https://nationalmap.gov/small_scale/printable/fedlands.html#list (accessed September 30, 2017).
5. FAR 1.601(a); FAR 1.602.
6. FAR 36.
7. General Services Administration website available at https://www.gsa.gov/real-estate/real-estate-services/for-businesses-seeking-opportunities/bidding-on-federal-construction-projects (accessed October 14, 2017).
8. FY 2016 Federal Real Property Profile (FRPP) Open Data Set, available for download at https://www.gsa.gov/policy-regulations/policy/real-property-policy/data-collection-and-reports/frpp-summary-report-library (accessed December 9, 2018).
9. JLL Office Outlook, 4Q 2016, United States, available at http://www.us.jll.com/united-states/en-us/Research/US-Office-Outlook-Q4-2016-JLL.pdf (accessed October 14, 2017).
10. JLL Office Outlook, 4Q 2016, United States, available at http://www.us.jll.com/united-states/en-us/Research/US-Office-Outlook-Q4-2016-JLL.pdf (accessed October 14, 2017).; "New York City's Office Market" report prepared by the Office of the State Deputy Comptroller for the City of New York, available at https://www.osc.state.ny.us/osdc/rpt10-2017.pdf (accessed October 14, 2017).
11. FY 2017 Federal Real Property Profile (FRPP) Open Data Set, available for download at https://www.gsa.gov/policy-regulations/policy/real-property-policy/data-collection-and-reports/frpp-summary-report-library (accessed February 10, 2019).
12. Fort Belvoir website http://www.belvoir.army.mil/history/Humphreys.asp (accessed October 14, 2017).
13. U.S. Army Corps of Engineers Fact Sheet as of February 2015, available for download at http://www.nab.usace.army.mil/portals/63/docs/FactSheets/FY15_Factsheets/VA_FtBelvoir.pdf; Fort Belvoir website: http://www.belvoir.army.mil/history/21C.asp (accessed October 14, 2017).
14. Fort Belvoir website: http://www.belvoir.army.mil/history/21C.asp (accessed October 14, 2017).
15. "BRAC at Fort Belvoir 1988–2011," written by Gustav Person, Installation Historian, U.S. Army Garrison Fort Belvoir, Virginia, available for download at http://www.usace.army.mil/Portals/2/docs/history/BRAC_History-Text.pdf (accessed October 14, 2017).
16. L. Berry, et al., 2004, "The Case for Better Buildings," Frontiers of Health Service Management, 21:1; Dewberry Design Charette Report, Community Hospital Fort Belvoir, March 1, 2007.
17. "BRAC at Fort Belvoir 1988–2011," written by Gustav Person, Installation Historian, U.S. Army Garrison Fort Belvoir, Virginia, available for download at http://www.usace.army.mil/Portals/2/docs/history/BRAC_History-Text.pdf (accessed October 13, 2017); US Army website article, available at https://www.army.mil/article/64690/fort_belvoir_community_hospital_opens (accessed October 13, 2017); Fort Belvoir Community Hospital website, "About Us," available at http://www.fbch.capmed.mil/about/history.aspx (accessed October 13, 2017).
18. "Belvoir to get millions in infrastructure improvements," Belvoir Eagle, July 23, 2009, available for download at http://www.belvoireagleonline.com/news/belvoir-to-get-millions-in-infrastructure-improvements/article_ad613b76-0dba-577d-b06a-89956f38555b.html (accessed October 13, 2017).
19. "Belvoir to get millions in infrastructure improvements," Belvoir Eagle, July 23, 2009, available for download at http://www.belvoireagleonline.com/news/belvoir-to-get-millions-in-infrastructure-improvements/article_ad613b76-0dba-577d-b06a-89956f38555b.html (accessed October 13, 2017); "BRAC at Fort Belvoir 1988–2011," written by Gustav Person, Installation Historian, U.S. Army Garrison Fort Belvoir, Virginia, available for download at http://www.usace.army.mil/Portals/2/docs/history/BRAC_History-Text.pdf (accessed October 13, 2017).
20. "BRAC at Fort Belvoir 1988–2011," written by Gustav Person, Installation Historian, U.S. Army Garrison Fort Belvoir, Virginia, available for download at http://www.usace.army.mil/Portals/2/docs/history/BRAC_History-Text.pdf (accessed October 13, 2017).
21. Public Law 110–161: Consolidated Appropriations Act, 2008, passed December 26, 2007; "BRAC at Fort Belvoir 1988–2011," written by Gustav Person, Installation Historian, U.S. Army Garrison Fort Belvoir, Virginia, available for download at http://www.usace.army.mil/Portals/2/docs/history/BRAC_History-Text.pdf (accessed October 14, 2017); "Military Base

Realignments and Closures: DOD Faces Challenges in Implementing Recommendations on Time and Is Not Consistently Updating Savings Estimates," Government Accountability Office, Report Number GAO-09-217, Published: January 30, 2009.

22. "Military Base Realignments and Closures: DOD Faces Challenges in Implementing Recommendations on Time and Is Not Consistently Updating Savings Estimates," Government Accountability Office, Report Number GAO-09-217, Published: Jan 30, 2009.

23. Department of Defense Directive Number 4270.5, "Military Construction," February 12, 2005.

24. UFC 4-010-01 DoD Minimum Antiterrorism Standards for Buildings, Chapter 1, 1-1 General.

25. Fort Belvoir Development Briefing for Southeast Fairfax Development Corporation by Colonel Mark Moffatt, Deputy Garrison Commander for Transformation and BRAC, Fort Belvoir, March 30, 2010, Unclassified; "Fairfax County Commercial Real Estate" presentation by Fairfax County Economic Development Authority, Curt Hoffman, Senior Manager of Real Estate Services, November 10, 2010, available at https://www.fairfaxcountyeda.org/sites/default/files/FCEDABRACUpdate1102010.pdf (accessed October 14, 2017).

26. Southeast Fairfax Development Corporation website, section on Development / Fort Belvoir, available at http://sfdc.org/fort-belvoir/ (accessed March 17, 2018).

27. "BRAC at Fort Belvoir 1988–2011," written by Gustav Person, Installation Historian, U.S. Army Garrison Fort Belvoir, Virginia, available for download at http://www.usace.army.mil/Portals/2/docs/history/BRAC_History-Text.pdf (accessed October 13, 2017).

28. "BRAC's Impact on Traffic Congestion and Quality of Life in Northern Virginia," Statement of the Honorable Jeffrey N. Shane, Under Secretary of Transportation for Policy, United States Department of Transportation Before the Committee on Government Reform United States House of Representatives Field Hearing on BRAC's Impact on Traffic Congestion and Quality of Life in Northern Virginia, August 31, 2006, available online at https://www.transportation.gov/content/bracs-impact-traffic-congestion-and-quality-life-northern-virginia (accessed March 17, 2018).

29. "BRAC Transit Service to Fort Belvoir," Finance & Administration Committee Information Item III-B July 7, 2011, Washington Metropolitan Area Transit Authority Board Action/Information Summary.

30. "BRAC at Fort Belvoir 1988–2011," written by Gustav Person, Installation Historian, U.S. Army Garrison Fort Belvoir, Virginia, available for download at http://www.usace.army.mil/Portals/2/docs/history/BRAC_History-Text.pdf (accessed October 13, 2017).

31. Virginia Department of Transportation press release, "First New Woodrow Wilson Bridge is Dedicated," May 18, 2006.

32. U.S. Department of Transportation, Federal Highway Administration website, TEA-21 Factsheet, available online at https://www.fhwa.dot.gov/tea21/factsheets/wwb.htm (accessed December 9, 2018); Public Law 106–346.

33. EPA website, section on National Environmental Policy Act Review Process, available at URL https://www.epa.gov/nepa/national-environmental-policy-act-review-process#EIS (accessed January 14, 2018).

CHAPTER 8

CASE STUDY—NATIONAL HARBOR

National Harbor is a visionary mixed-use development located on approximately 340 acres along the Potomac River in Maryland just south of Washington, D.C. The project's remarkable scope and daring location make it unique within the D.C.-Virginia-Maryland metro area. At the time of its inception in 1995, there was virtually no commercial development in the surrounding area and several previous efforts to develop the site had failed. However, after more than a decade of tenacious effort by the project's developer, the Peterson Companies, National Harbor's first phase opened along the Potomac waterfront in 2008. The waterfront development successfully delivered a critical mass of uses that created an instant sense of place and tapped unmet demand in the area. National Harbor has continued to grow since that initial delivery and currently includes a vibrant and well-curated mix of uses including office, hotel, retail, restaurant, residential, entertainment, a marina, and conference venues. A dedicated, internal bus service shuttles visitors and residents between the project's different areas, including an outlet mall, a casino, and a 180-foot Ferris wheel on the waterfront. With an estimated resident population of approximately 3,800, more than 240 on-site employers, and attracting an estimated 11 million visitors annually, National Harbor and its vicinity are recognized by the U.S. Census Bureau as a census designated place (CDP) on due to its magnitude.[1] It is truly a substantial achievement in development.

However, despite its current success, the concept behind National Harbor was less obvious at the beginning of the development process. The sheer scope of the project presented its own set of challenges, along with other difficulties related to zoning, political and economic conditions, and environmental and cultural issues. Overcoming these and other hurdles effectively required the development to be underway for more than a decade before any rental or sales revenue was achieved. The following sections of this case study examine the National Harbor project in greater detail and illustrate the commitment that was necessary to bring the project vision to reality.

8.1. BACKGROUND

National Harbor is part of the Henson Creek-South Potomac Planning Area located in Prince George's County Maryland. The site features 1.25 miles of prime waterfront, including beach areas, on the banks of the Potomac River, as shown in Fig. 8.1. It is further bounded by the Capital Beltway (I-495/95) to the North and the residential community of Oxon Hill, comprised largely of low-density single-family residential homes, to the East and South. Portions of the site abut land

FIGURE 8.1   National Harbor waterfront. (*Courtesy of the Peterson Companies; photo credit: Rich Martin.*)

owned by the Maryland-National Capital Park and Planning Commission (M-NCPPC), including the Betty Blume Neighborhood Park and the grounds of the historic Oxon Hill Manor. The overall property is effectively divided into three sections that were assembled through multiple separate acquisitions: the "Waterfront Parcel" of 224.85 acres, the "Beltway Parcel" of 64.62 acres, and the "Tanger Parcel" of 50 acres. While the Tanger Parcel was not a part of the original National Harbor land acquisition, it has since been incorporated into the greater development and will be treated as part of National Harbor for purposes of discussion herein. Note that the project also includes approximately 246 acres located beneath the Potomac River that forms the basis for the piers and marina, as shown in Fig. 8.2.

The availability of a vehicular entrance from the Capital Beltway (I-495/95) means that the site is directly accessible by well over 230,000 cars that drive past every day.[2] Located in close proximity to both downtown Washington, D.C. and Old Town Alexandria in Virginia, the site is near to the some of the most affluent population centers in the nation, as shown in Fig. 8.3. Lack of competing properties in the surrounding area make the project a natural draw for local Maryland residents as well. In addition to locational benefits, the site has an abundance of natural features, such as ideal topography on the Waterfront Parcel featuring a constant slope of 5 percent rising from the river

up to the back of the site, creating the potential for tiered buildings with stunning views.

Given its overall appeal, the site, historically known as Smoot Bay, understandably attracted previous development efforts. The first concerted effort was a project known as Bay of Americas that failed to evolve past the initial stages of zoning approvals. The Bay of Americas site, which consisted of the Waterfront and Beltway sites as separate parcels, was subsequently purchased by Virginia developer James Lewis. Lewis named his project PortAmerica and pursued its development throughout the 1980s and early 1990s. His vision featured a residential community with a mixed-use component on the Waterfront Parcel and a 52-story office tower located on the Beltway Parcel.[3] The design of the office component, which was intended to be designated part of the World Trade Centers Association, envisioned a significant increase in height and density when compared to the surrounding low-density residential uses. Controversy surrounded the office building, including a reported conflict with the flight path from nearby Reagan National Airport, and ultimately resulted in a reduction of the proposed building height.[4] A subsequent redesign created a World Trade Center complex consisting of a central 23-story office "tower" with six shorter surrounding office buildings. Despite notable support for the project, success

FIGURE 8.2 National Harbor property boundaries. (*Courtesy of the Peterson Companies.*)

in obtaining necessary permits, and the transfer from the government of a key tract of land to connect the Waterfront and Beltway Parcels, the economic conditions of the early 1990s forced major investors to withdraw from the project at a loss and led to the developer's eventual default on project loans.[5] The savings and loan (S&L) crisis of the late 1980s and early 1990s landed the defaulted PortAmerica loan and the issuing S&L institution into conservatorship with the federally run Resolution Trust Corporation (RTC), which was tasked with liquidating the assets of insolvent entities (among other things).[6] Regardless of the PortAmerica team's efforts to repurchase the property, the RTC ultimately put the land up for bid.

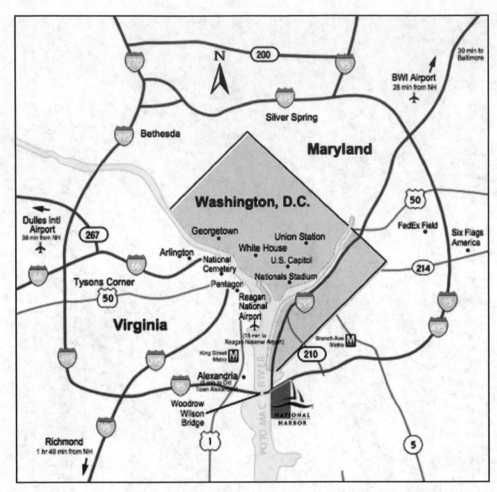

FIGURE 8.3   Location map of National Harbor. (*Courtesy of the Peterson Companies.*)

Though unfortunate in its own right, the demise of the PortAmerica project helped create the opportunity for National Harbor. The RTC's efforts to solicit bids for the property attracted the attention of the Peterson Companies' executives, who visited the site in the winter of 1995. It was the Waterfront Parcel that first inspired the Peterson Companies founder Milt Peterson's vision for National Harbor. Struck by a combination of possibility and patriotism roused by riverfront views toward the nation's capital, Peterson knew he had found his next big project. Turning the failed site into a success would also become the greatest challenge he had faced in his career as a developer. A bid was promptly submitted citing the Peterson Companies' previous experience completing bids with RTC as well as the firm's expertise with other large-scale development projects in the region, such as Fair

Lakes and Burke Centre. The firm closed on the 224.85-acre (above ground acreage) Waterfront Parcel in February 1996 and on the Beltway Parcel in December of the same year. Over the next several years, the Peterson Companies proceeded to acquire additional property to create the greater project site across the Waterfront, Beltway, and Tanger Parcels. For the project team dedicated to National Harbor, this was the beginning of what would become a decades-long endeavor.

The internal project team was guided by Peterson's mantra that "success is the degree to which you meet your potential," a principle that applies not only to personal and professional development, but also to the development of sites. Peterson maintained that a truly great site deserves a great development and to pursue anything less at National Harbor would have been irresponsible.

A grand scope was also necessary given the site's location: with no existing density or commercial and retail draws, the project needed a big concept in order to be successful. Unlike phased urban and peri-urban projects, National Harbor could not be created incrementally because there was no existing market in the surrounding area to drive business or attract patrons while future phases were under construction. Rather, the development team would have to create something compelling enough on opening day to draw people from their familiar markets in D.C., Maryland, and Virginia. Further, the project not only had to appeal to shoppers as a destination, but also had to convince the retailers themselves that the seemingly isolated location was viable. The same was true for office tenants, hoteliers, and potential residents. To survive, National Harbor would have to deliver a critical mass of mixed uses on the day it opened.

Peterson and his team were committed in their efforts to develop a superior concept for the site: the team held internal charrettes to inspire creativity, building off each other's ideas as they were driven by Peterson to imagine everything the site could become; key team members travelled to Europe to study inspirational places, such as the Las Ramblas promenade in Barcelona and the Spanish Steps in Rome; and specialist consultants were recruited from around the country for input, including renowned landscape architect Stuart O. Dawson and lead Disney Imagineer, Bob Weiss. A vision for National Harbor began to emerge. It would bring scale and economic resources to the county while embracing a flexible concept that could evolve to meet changing demands and needs over time.

Despite the renewed enthusiasm from a new developer, many of the historical problems that plagued the site as PortAmerica still existed. For example, although the site could be connected to the Capital Beltway (I-495/95), the traffic congestion and lagging redevelopment of the adjacent Woodrow Wilson Bridge were problematic. The existing entitlement process was laborious and poorly able to support the flexibility required by a project with the scope of National Harbor. Further, there were a range of environmental and cultural concerns resulting from the site's location on the Potomac River and on the grounds of the historic Oxon Hill and Salubria plantations, which included a family cemetery. In addition to inherited challenges, National Harbor also encountered its own set of difficulties. The development timeline was so protracted that the project incurred 10 years of entitlement, development, and construction expenses before it generated any revenue. Complicating matters, the first phase of the development opened in April 2008 on the precipice of the economic tsunami that was the "Great Recession" in the United States, creating financial uncertainty that had to be resolved between the Peterson Companies and the consortium of six banks providing funding for the project. These and many other challenges were addressed and overcome in the 12-year long development journey between the purchase of the site and the opening of National Harbor.

8.2. PROJECT TIMELINE

The National Harbor project was a monumental undertaking that continued well beyond the delivery of its initial phase. As such, a comprehensive list of all project dates would be prohibitive. The following presents a timeline of some of the most significant project milestones, many of which were heavily interrelated and required long lead times before advancing the development.

Winter 1995—The Peterson Companies' executives and development team visit the site.

February 1996—The Waterfront Parcel is purchased via submission of winning bid to RTC.

December 1996—The Beltway Parcel is purchased from Signet Bank's REO department.

1997 to 1998—The original U.S. Army Corps of Engineers wetland permit for PortAmerica is modified and extended to accommodate plans for National Harbor (several additional modifications will occur over the following decades as the project grows).

1997—Site clearing and grading begins on the Waterfront Parcel.

June 17, 1997—Prince George's County Council approves various legislation that enables the National Harbor concept, including the creation of a "Waterfront Entertainment/Retail Complex" in the zoning ordinance.

February 11, 1998—National Harbor is approved as a Waterfront Entertainment/ Retail Complex for zoning purposes.

June 10, 1998—Approval of conceptual site plan (CSP) for National Harbor (with conditions).

1999—Jurisdiction for project approvals is transferred by act of Congress.

2000 to 2001—Various critical site studies are undertaken and related approvals obtained, including 100-year flood plain study, tree conservation planning, and erosion control planning.

July 2001—Preliminary plan of subdivision (PPS) approved (with conditions).

November 2003—Revised PPS approved by Planning Board.

2003 to 2005—Roads are constructed on the Beltway Parcel.

December 2004—First Tax Increment Financing ("TIF") bond funding for public infrastructure and groundbreaking event for the project.

February 2005—Purchase and sale agreement is finalized with Gaylord Entertainment for 42 acres for the development of a nearly 2,000-room resort hotel and convention center.

September 2005—Gaylord National Resort and Convention Center begins construction; residential uses permitted at National Harbor.

April 2006—National Harbor retains its M-X-T zoning in the Approved Master Plan and Sectional Map Amendment for the Henson Creek-South Potomac Planning Area.

June 2006—The "outer loop" of the new Woodrow Wilson bridge opens, allowing improved access to the site.

July 2006—Construction begins on 15 different buildings/lots on the Waterfront Parcel, necessitating an elaborate construction staging and logistics plan; construction prices are increasing significantly.

April 2007—The Peterson Companies secures construction financing from a consortium of six banks lead by Bank of America.

June 2007—Residential sales office opens and condominium units are nearly 100 percent pre-sold within two months.

December 2007—The "Great Recession" begins in the United States and buyers begin withdrawing from their purchase commitments for condominium units.

2008—Slot machine gambling is legalized in Maryland.

April 1, 2008—Official opening of the nearly 2,000-room Gaylord National Resort and Convention Center and the National Harbor Waterfront Parcel first phase, consisting of: 200,000 square feet (sf) office, 250,000 sf of retail/restaurant/ entertainment space, 423 condos, 950 additional hotel rooms, and 2 piers.

June 2008—The "inner loop" of the new Woodrow Wilson bridge opens, completing site access.

Fall 2008—The full force of the Great Recession begins to impact the economy; a large majority of residential condominium buyers withdraw from their sales contracts and the units have to be remarketed for sale.

May 2009—Walt Disney Co. buys 11 acres along the ridge near the top and rear of the Waterfront Parcel with views of National Harbor's waterfront; Disney plans a 500-room resort.

May 2011—A joint venture is formalized for the development of a Tanger Outlet shopping center at National Harbor.

November 2011—Disney rescinds decision to participate in National Harbor; the Peterson Companies ultimately buys back the lot.

April 2012—Application is filed for approval of Tanger Outlets.

October 2012—DSP Approval received for Tanger Outlets.

November 2012—Groundbreaking for Tanger Outlets; a special referendum allows a gaming facilities in Maryland casinos and grants approval for a 6th venue to be built in Prince George's County.

April 2013—MGM enters into ground lease for 23 acres on the Beltway Parcel.

November 2013—Tanger Outlets opens.

February 10, 2014—Application is filed for approval of the MGM resort and casino.

May 8, 2014—Board hearing date for MGM.

May 23, 2014—The Capital Wheel opens.

July 2014—Prince George's County Council approves MGM; site work begins shortly thereafter.

November 24, 2015—Plans submitted for approval on an additional 1.278 million sf of mixed-use development on the Beltway Parcel.

March 17, 2016—Plans are approved (with conditions) for an additional 1.278 million sf on the Beltway Parcel.

December 2016—MGM resort and casino opens.

After more than 20 years of development, improvements to the overall project are still underway with plans for work to continue beyond 2020. National Harbor was envisioned to be a flexible project that could evolve to meet future needs and it continues to exceed its potential in this regard with each new addition.

8.3. PROJECT COMPONENTS AND COSTS

National Harbor has approvals in place for a total of 7.341 million sf of a master planned mixed-use community including 2,500 residential units. At completion, the project will feature over 800,000 sf of office alongside more than 1.7 million sf of retail, restaurant, and entertainment venues paired with 2,666 residential units and more than 4,500 hotel rooms. At its current level of build-out, National Harbor represents approximately $4 billion in total project costs distributed between numerous different components, including site infrastructure. Thus far, the project has realized 4.2 million sf of commercial development across multiple product types, 3,246 hotel/timeshare rooms in seven hotel brands, and another 2.1 million sf of residential development. The Peterson Companies is the developer-owner of much of the project but also acts as master developer, land-lessor, and partner under various arrangements in which different parcels are sold, ground leased, or held under a joint venture (JV). While specific individual cost break-outs would be too complex to list in their entirety, some of the major project cost components are:

- MGM resort and casino: $1.4 billion
- Gaylord hotel and convention center: $1.2 billion
- National Harbor PH IA: $900,000 million
- Internal road and infrastructure improvements: $90 to $100 million
- Tanger Outlets: $100 million
- Water pumping station: $15 million
- Waterfront and Marina improvements: $30 million

The rest of this section provides an overview of the different product types included in the development of National Harbor, as shown in Fig. 8.4.

8.3.1. Retail

The project features approximately 720,000 sf of retail/dining/entertainment (RDE) space as of 2018, including shopping, restaurant, and performance venues, as summarized in Fig. 8.5. The retail component is most densely concentrated in the Tanger Outlets, operated by a public real estate investment trust (REIT) that maintains a 50 percent ownership share of the 80-store shopping outlet at

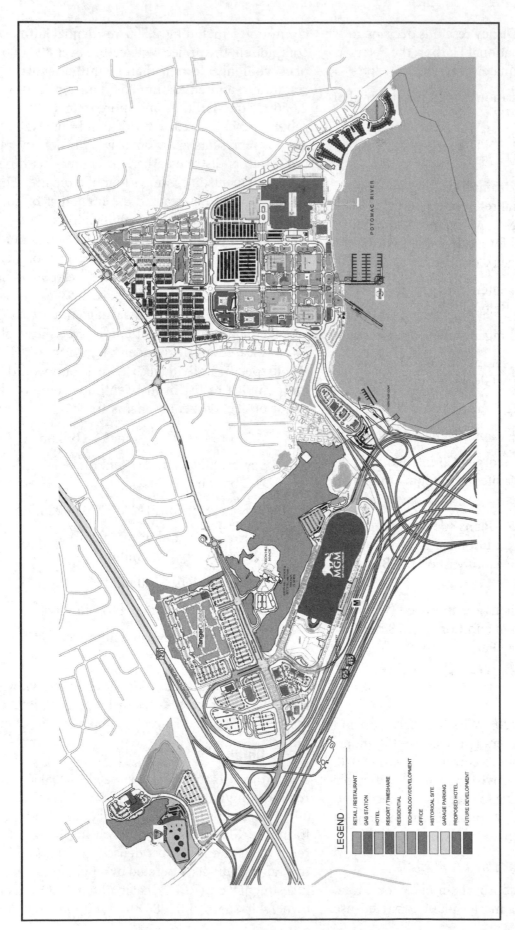

FIGURE 8.4 National Harbor detailed site plan. (*Courtesy of the Peterson Companies.*)

Retail Summary
Total: 722,584 sf
• Including 160 shops and 40 restaurants
Waterfront parcel: 326,733 sf
Tanger Outlets: 341,156 sf
MGM: 54,695 sf

FIGURE 8.5   Retail summary.

National Harbor through a joint venture with the Peterson Companies.[7] The outlet stores brought a reported 1,000 jobs and an estimated $6.5 million in annual tax revenue to Prince George's County.[7] Retail is also an important part of the Waterfront Parcel, where street-level boutique style space and restaurants enhance the National Harbor experience. A further selection of upscale stores and restaurants is offered on the Beltway Parcel at the MGM National Harbor resort and casino. Traditional retail and dinning entertainment is supplemented by a dedicated year-round event calendar featuring innovative programing, such as the Sunset Concert Series, viewing parties for major professional sporting events, free public exercise classes, a farmer's market, outdoor movie nights, seasonal festivals, and the Gaylord's ice sculpture exhibit in winter.

8.3.2. Hospitality and Convention

In 2017, Washington, D.C. hosted 20.8 million domestic tourists and benefitted from $7.5 billion in visitor spending.[8] National Harbor captures its share of the area's tourism with seven hotel and timeshares encompassing a total of 3,246 rooms plus convention and event space, as outlined in Fig. 8.6. The development will boast more than 4,500 hotel/timeshare rooms at final build-out. Not only can out-of-town visitors enjoy the offerings at National Harbor, but guests are connected to D.C. and Virginia via a water taxi and bus lines, which allow them access a myriad of D.C. sites, including the Smithsonian museums, national monuments, and other historic sites.

One of the major attractions drawing tourists to National Harbor is the MGM National Harbor resort and casino and its entertainment program. The casino property, which received the last of six gambling venue licenses from the

Hospitality Summary
Total hotel rooms: 3,246
Gaylord National Resort & Convention Center
• 1,996 rooms
• 546,889 sf of convention center & meeting space
MGM National Harbor
• 308 rooms
• 620,000 sf casino and entertainment space
• 50,000 sf convention area
• 1,200-seat theater
AC Hotel National Harbor By Marriott
• 184 rooms
Hampton Inn & Suites
• 151 rooms
Residence Inn by Marriott
• 162 rooms
Westin
• 195 rooms
Wyndham Vacation Resorts at National Harbor
• 250 timeshare rooms

FIGURE 8.6   Hospitality summary.

state of Maryland, was developed on a 23-acre ground lease.[9] Since its opening in 2016, MGM has brought 4,000 jobs and an estimated $40 to $45 million in tax revenue to Prince George's County.[10] The project received certification as a LEED Gold property in 2017.[11]

Business and conference travelers at National Harbor are served by both MGM and the extensive convention space at the Gaylord National Resort & Convention Center. The nearly 2,000-room Gaylord National, shown in Fig. 8.7, sits on 42 acres of land purchased from the Peterson Companies and was the catalyst that drove the completion schedule for much of the first phase of Waterfront Parcel completed in 2008. The Gaylord hotel brand was acquired by Marriott International in 2012.[12]

8.3.3. Residential

Plans for National Harbor forecast a total of 2,666 residential units when build-out is complete. As of 2018, there are already more than 1,400 existing residential units in-place representing over 2 million sf in a mix of condominiums, townhouses, and apartments, as shown in Fig. 8.8. The Peterson Companies developed the condo and apartment buildings and the townhome lots, while the townhomes themselves were developed

FIGURE 8.7   Gaylord National Resort & Convention Center. (*Photo credit: Gaylord National Resort.*)

by Integrity Homes. Pricing for each type of home is reflective of views and river access; proximity to D.C.; high-level finishes and amenity packages; and, of course, the convenience and distinction of being part of National Harbor itself. Apartments and condo units are situated in the town center portion of the Waterfront Parcel with townhomes located toward the upper, rear portion of the site away from the busy retail and entertainment areas. Thanks in part to the stepped site design, westward-facing residential units in buildings farther back in the development still have views of the marina and Potomac River.

8.3.4. Office

With approximately 292,000 sf of existing office space, National Harbor has only reached approximately 35 percent of its anticipated total office build-out. Most current office space is located on the Waterfront Parcel in buildings with small to medium sized footprints, as shown in Fig. 8.9. The approachable floor sizes allow smaller firms to enhance their corporate identities as full-floor users.

Residential Summary
Total residential units: 1,406
Condominiums
• 248 units at The Haven at National Harbor
• 163 units at Fleet Street
• 242 units at One National Harbor
• 18 units at Waterfront
Townhomes
• 473 units at Uplands
Apartments
• 262 apartments at The Esplanade at National Harbor

FIGURE 8.8   Residential summary.

Office Summary
Total rentable Class A space: 456,352 sf
• 1 National Harbor (Bldg L): 9,023 sf
• Building B: 23,263 sf
• Building C: 14,125 sf
• Building E: 10,583 sf
• Building JW: 60,708 sf
• Building M: 60,840 sf
• Building P: 11,298 sf
• Building Q: 19,403 sf
• Parcel 4a MGM: 29,109 sf
• Parcel S2 Medical Pavilion: 93,000 sf
• Oxon Hill office building: 125,000 sf

FIGURE 8.9   Office summary.

8.3.5. Public Art

Well-designed streetscapes and architecture are not the only elements that add to the project's overall appeal. The environment at National Harbor is further enhanced by an actively curated collection of public art consisting of 20 works including mosaics, sculptures, and large-scale installations, including that shown in Fig. 8.10. Among the more recognizable are:

- "The Awakening" by artist J. Seward Johnson, an installation featuring a giant emerging from the earth (moved from Hains Point in D.C. where it was originally installed in 1980).

- "The Beckoning" an abstract work by sculptor Albert Paley.

- "Eagles" by sculptor Albert Paley featuring two stainless-steel eagles "flying" 65 feet over the main promenade stairway leading to the waterfront.

- Numerous sculptures by Ivan Schwartz of Studio EIS including George Washington, Abraham Lincoln, Frederick Douglass, and others.

8.4. PROJECT CHALLENGES

Given its size and lengthy timeframe, it is unsurprising that the project team encountered a wide range of challenges throughout the development of National Harbor. The following discussion outlines three of the more notable challenges the project faced as well as the efforts required to bring about their successful resolutions.

8.4.1. Challenge #1: Zoning

The site's existing zoning in conjunction with the approval process of the Maryland-National Capital Park and Planning Commission (M-NCPPC), were immediate challenges for National Harbor. As is true in many cases around the country, existing master plans are often decades old and tied to Euclidean

FIGURE 8.10 Public art at National Harbor. (*Courtesy of the Peterson Companies.*)

zoning ordinances, neither of which is especially conducive to rapidly urbanizing areas or a contemporary ideas about space and land use. Although Prince George's County zoning ordinances did include mixed-use zones, different portions of the site fell under the "Mixed Use—Transportation Oriented" (M-X-T), "Rural Residential" (R-R), and "Residential Medium Development" (R-M) zoning designations, each of which had various restrictions on uses and density. Further, each was subject to an entitlement process that required multiple, lengthy stages of submission, review, and public hearings before any approvals could be received. Specifically, this included the requirement for a separate submission and approval of a CSP, a detailed site plan (DSP), and a PPS.

Rather than be limited by these constraints, the team followed Peterson's philosophy that conceiving a great development does not begin with a review of what is permitted by existing zoning, but rather by exploring the site's maximum potential and then pursuing changes in zoning as necessary. The team's visioning exercise for National Harbor was extensive and incorporated as many expert views as possible. In addition to the internal charrettes and international inspiration-gathering travel previously mentioned, the team went so far as to produce a white paper written from the perspective of the site itself to explore what it "wanted" to become.

Once the vision of National Harbor as a grand, mixed-use complement to Washington, D.C. had emerged, it became obvious to Peterson and the team that significant flexibility would be needed in order to bring the overall plan to fruition. The team knew the project would take decades to complete and that working in the late 1990s, it would be impossible for them to accurately predict building-specific needs as many as 30 or more years into the future as the project developed throughout the 2000s, 2010s, 2020s, and beyond. It was critical that the National Harbor plan be able to evolve in order to respond to changing market conditions and consumer demand over time. The team's prediction has certainly held true given that the technology, lifestyles, and space-use of the 1990s were radically different than today's environment, which itself continues to evolve with the rise of autonomous vehicles, smart cities, and other technological developments. However, the existing entitlement process in Prince George's County in the late 1990s meant the team had only two choices, either (1) face the tedious process of submitting each portion of the project individually and waiting for more than a year to move through planning approval before each new element of National Harbor could be realized or (2) commit to a single vision and be rigidly held to its approval, unable to modify key aspects of the project in the future, such as building heights, locations, and uses in response to changing needs. Neither was an appropriate option for a scheme as large and complex as National Harbor.

The project was essentially undevelopable if the team were to be subject to the normal entitlement process and procedures for the property. Yet, undaunted, Peterson pressed forward. The first step to "liberating" National Harbor was to remove federal government project oversight by the National Capital Planning Commission and bring it under the authority of those with a more directly vested interest in the project's success: Prince George's County and the M-NCPPC. Achieving this took an Act of Congress to relieve federal oversight, which was granted in 1999 after several years of effort from political supporters as well as significant lobbying efforts.[13] However, the existing zoning still was not flexible enough for the project's needs.

The second step was to seek a change in zoning ordinances to help increase flexibility and development response times. The National Harbor team worked with all levels of government, including the Maryland State Senate, Governor's Office, and the Prince George's County Executive, to explain the merits of the project and the benefits to the county. National Harbor would give Prince George's County a prominent destination that it had never before had as well as create jobs and new homes. Importantly, it would also be a substantial source of tax revenue for the county. Studies completed at the time demonstrated that the project could generate an estimated $1.8 million in annual taxes, a stunning amount for the county. Many officials who had witnessed the downfall of PortAmerica understood the need to support National Harbor in order for the project to be viable and bring the various benefits to the county. To achieve this, a zoning change was proposed to create a new designation called the "Waterfront Entertainment/Retail Complex" which could exist under the M-X-T and other zoning categories. A crucial element of the designation was the potential to submit a CSP while waiving the requirement for a DSP for Waterfront Entertainment/Retail Complex projects. This allowed the Planning Board and County Council to retain authority over the approval of a general plan while empowering the developer to refine the specifics as needed over time without returning for additional approvals.

While there was certainly resistance to the plans for National Harbor, both in government and from citizen groups, the undeniable benefits ultimately outweighed concerns and in 1997 a Zoning Bill was passed creating the new "Waterfront Entertainment/Retail Complex" use designation, as described in Fig. 8.11. The original ordinance, passed by Council Bill 44-1997, would undergo amendments through two subsequent Bills in 2005 and 2011. The team simultaneously pursued a rezoning to designate the Waterfront Parcel as a "Waterfront Entertainment/Retail Complex," which was granted upon County Council approval of Resolution Nos. 44-49.

The third and final step to overcoming the zoning challenge was to obtain the necessary entitlements for the project under the new zoning scheme. It took the team a full year to create a CSP for submittal on the National Harbor Waterfront Parcel. Given that a waiver of the DSP was sought, the CSP needed to include greater detail than a typical submission. Even with the new flexibility, this remained a difficult task because project details were still being contemplated in the absence of any tenants or users. Nonetheless, the team was successful in proposing a plan that "outlined general arrangement of uses of lots, the circulation system, and those areas of the site to be conserved,"[14] as required. It was malleable enough to adapt when future specifics became known, yet still able to including certain setbacks, a grid system for different future blocks of development, and demonstrate a natural growth in height and density toward the center of the site. The plan attempted to be sensitive to neighboring properties through the use of landscaping and other buffers. The CSP was ultimately successful in moving through the approval process at all levels, including the County's District Council, which had review authority of M-NCPPC's decision, and was approved in 1998 as one of the first examples of a "form-based" development in the county. With the first approval secured, the team began working on the PPS and the processes of passing adequate public facilities tests, which were mostly focused on traffic-related issues. Studies were also conducted throughout the late 1990s and early 2000s and a host of necessary permits were sought and obtained. The final plan of subdivision was approved in November 2003 after required revisions were made to the original submission. After eight years of work, the development of National Harbor could finally begin.

Subtitle 27 - Zoning
Part 2 - General
Division 1 - Definitions
Section 27-107.01. - Definitions

(256.3) Waterfront Entertainment/Retail Complex:
A contiguous land assemblage, no less than twenty-five (25) acres, fronting on the Potomac River, and developed with an array of commercial, lodging, residential, recreational, entertainment, social, cultural, or similar uses which are interrelated by one (1) or more themes. A gas station located within a Waterfront Entertainment/Retail Complex may include a car wash as an accessory use, provided the car wash is within or is part of the building(s) for which design and architecture are approved in the Detailed Site Plan for the gas station.

(CB-44-1997; CB-20-2005; CB-22-2011)

FIGURE 8.11   Waterfront Entertainment/Retail Complex zoning description.

8.4.2. Challenge #2: Project Scope

With approvals for a total of 7.341 million sf, the size of the National Harbor itself created challenges even with development phased over the course of several years. The delivery of the Waterfront Parcel, the first phase to be completed, was a monumental undertaking in its own right. The initial opening featured the simultaneous delivery of the Gaylord National Resort & Convention Center along with 17 other buildings encompassing: 200,000 sf of office space; 250,000 sf of retail, dining, and entertainment space; nearly 3,000 hotel rooms; 423 condominium units; a public plaza; and two piers. Achieving such a feat required obtaining significant funding, finding retail tenants willing to commit to the as yet untested concept of a large stand-alone development in Prince George's County, and also involved the coordination and collaboration of an extensive project team. The list of National Harbor contributors is extensive including participation by local and minority-owned firms and subcontractors. A nonexhaustive representation of technical project team members is shown in Fig. 8.12. Note that this list does not include the many legal and business team members that contributed to the success of National Harbor.

The firms listed in Fig. 8.12 represent the team for the first phase of the Waterfront Parcel; subsequent portions of the project, such as the MGM resort and casino, had their own teams.

The challenge of simultaneous building on the scale of National Harbor can perhaps be best appreciated through the lens of construction management as multiple individual teams worked to essentially create an entire mini city from the ground-up. The project scope was so large and the site so compact that full time construction logistics and security personnel were hired to manage affairs on-site. At one time the project site had as many as 28 active construction cranes and hosted 5,000 workers per day, making it potentially the largest single construction site in the country. Special planning was required to assign a horizontal swing elevation to each tower crane to make sure that each could reach the necessary construction height required for building while also being able to swing freely for 360 degrees (for work or in the wind) without hitting other cranes. In addition to crane logistics, the close proximity of the 15 different building work sites needed to be managed. Each construction contractor had a separate staging area fenced off, some as large as 15,000 sf, to create a series of individual compounds. This allowed each firm to maintain control of their own materials, equipment, and on-site facilities. The complexity of this coordination was captured in detailed graphics and provided to the construction team members, as shown in Fig. 8.13. Parking fields were located at the top of the hill on the Waterfront Parcel with dedicated parking areas color coded for the use of different firms. Finally, a "Rules and Regulations" pamphlet was issued to help reduce the friction that naturally occurred on such a busy, compact site.

Adding to the pressures of construction, the agreement with Gaylord required almost 200,000 sf of retail space to be delivered, leased, and operating by the time of the hotel's grand opening. This was an imperative from Gaylord's perspective given the site's isolated location. As described previously, critical mass at National Harbor could not be delivered incrementally due to the lack of surrounding density; there were no other operating business in the immediate area to attract patrons. The Gaylord contract imposed stiff penalties for failure to perform and this meant that not only did the surrounding buildings need to be completed in time, but also that the Peterson Companies leasing team had to attract retail tenants to the vision of National Harbor before the site was ready. Leasing efforts were under way before there were even roads in place on the site; yet many retailers wanted to wait until after Gaylord opened, which would have required the road infrastructure to be completed, before committing to the project.

In addition to on-site circulation, successfully creating access from the adjacent Capital Beltway (I-495/95) was critical for the project. The Dewberry Transportation group provided design solutions for access into National Harbor. The design of the Woodrow Wilson Bridge replacement was underway and the National Harbor access needed to be coordinated with both the beltway interchange with I-295 and Oxon Hill Road (MD 210). The complexity of the interchange at I-495/I-295/National Harbor was increased by the need to provide an additional "leg" to the interchange in order

Planning & Design Consultants
1. Development Design Group - Master Architect
2. Land Design, Inc.- Landscape Architecture
3. Sasaki & Associates - Master Landscape Architects & Environmental Graphics
4. Streetsense Retail Advisors - Master Planner
5. Wetlands Studies & Solutions - Wetlands
6. Wells & Associates - Traffic & Parking

Architectural Design Firms
1. Braun & Steidl Architects - Marriott Residence Inn (Bldg H2)
2. Brown Craig Turner Architects - Buildings B, C, E, H1
3. Campbell Architects- Westin (Bldg A)
4. Davis Carter Scott Design - Building D
5. Desman & Associates - Garage/Retail Building P
6. Gensler Associates - Gaylord National Resort & Convention Center
7. Hickock Cole Architects - Office/Retail/Garage (Bldg J & M)
8. Moffat & Nichol Engineers - Marina/Piers
9. Sasaki & Associates - Public Plaza & American Way
10. STV Architects - Building O (Hampton Inn & Suites)
11. WDG Architecture - Buildings K, L, & Q (Wyndham)

Engineering Consultants
1. Delon Hampton & Associates – Bridge
2. Dewberry
3. ECS LTD, Inc
4. E.K. Fox & Associates - Electrical
5. Engineering Consulting Services – Geotechnical, Materials
6. Hillis-Carnes Engineering - Materials Testing & Inspections
7. Schnabel Engineering Associates – Geotechnical
8. Sheladia Associates, Inc. – Inspections, Plan review
9. Soltesz, Inc
10. Testing & Inspections
11. Traffic Group
12. Watek Engineering – Pump Station

Contractors
1. Aggregate Industries
2. Arnold Parreco & Sons, Inc.
3. Balfour Beatty Construction
4. Cherry Hill Construction
5. Cianbro Corporation
6. Coakley & Williams Construction, Inc
7. Coastal Design & Construction
8. Electrosonics
9. Facchina-McGaughan, LLC
10. Herman/Stewart Construction
11. Hitt Contracting, Inc.
12. John E. Kelly & Sons
13. K.W. Miller, Inc.
14. M.C. Dean, Inc.
15. NORAIR Engineering Corp.
16. Perini Tompkins JV (Gaylord National)
17. Proctor-Denison Landscaping & Trucking
18. Shirley Contracting, Inc.
19. Southern Maryland Dredging
20. The Clark Construction Group and Clark Multi-Family Builders
21. The Whiting-Turner Contracting Company
22. Total Engineering, Inc.
23. W.F. Wilson & Sons
24. Worthington Contracting, Inc.

FIGURE 8.12   National Harbor technical project team members.

to provide access to the Beltway Parcel. In coordination with the Peterson Companies and other consultants including Soltesz (National Harbor's on-site civil engineer) and JMT (the 'section' designer for the Woodrow Wilson Bridge interchange), Dewberry developed a conceptual layout that accommodated all connections into National Harbor while minimizing impacts to the on-going

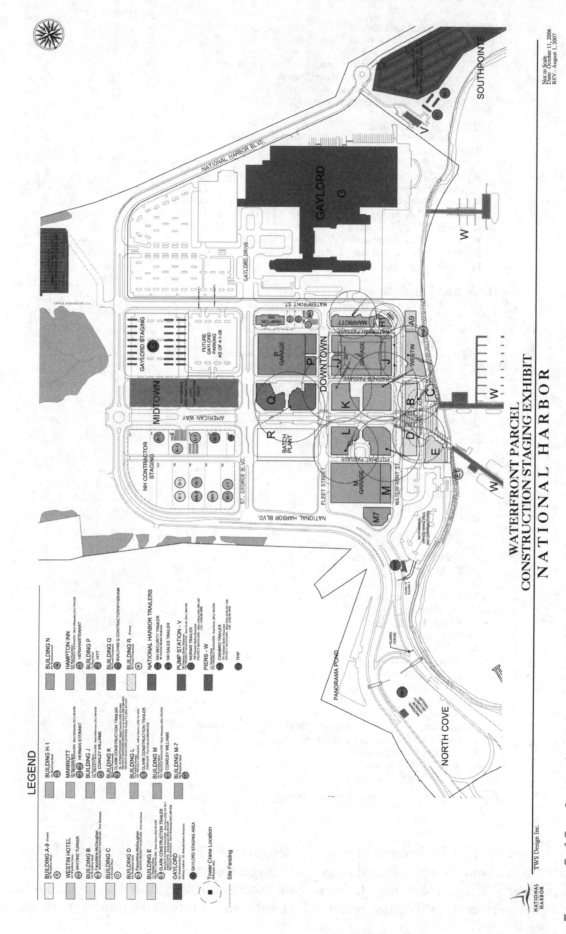

FIGURE 8.13 Construction staging and crane logistics plan.

Woodrow Wilson Bridge Replacement design. In a series of meetings and reviews with the Maryland State Highway Administration (MSHA) and the Federal Highway Administration (FHWA) over a 6-month period, Dewberry was able to gain approvals of the proposed access plan, ensuring that the Peterson Companies had access to both the Waterfront and Beltway Parcels, overcoming a significant hurdle in the effort to develop National Harbor.

8.4.3. Challenge #3: Environmental and Cultural

By virtue of changing the natural landscape, development projects often encounter environmentally and culturally sensitive conditions. Urban redevelopment projects, which often excavate to deeper levels than prior uses, may uncover previously undisturbed human remains, ancient structures, boats, and a range of other archeologically relevant finds. Circumstances are no less certain for greenfield developments, which may be located on historic sites or are located near sensitive bodies of water and face additional environmental scrutiny. National Harbor encountered both environmental and cultural challenges in the course of the project's many years of development. These ranged from the need to accommodate a pair of protected nesting bald eagles to concerns over the capacity of the newly built pumping station. As is often the case for developments in environmentally sensitive areas, National Harbor also faced its share of citizen opposition and legal challenges by conservation groups, many of which were also opposed to the redevelopment of the Woodrow Wilson Bridge.

With frontage along the Potomac River, the National Harbor site is part of the greater Chesapeake Bay watershed and ecosystem. This means that, in addition to typical wetlands concerns, the site is also part of a tidal network. As such, development of the site was subject to the Chesapeake Bay Area Protection Act enacted by the Maryland General Assembly in 1984. The language of the Act is embedded in the State of Maryland's annotated code under natural resources in Title 8 Water, Subtitle 18 Chesapeake and Atlantic Coastal Bays Critical Area Protection Program. The law defines the "critical area" as all land within 1,000 feet of the high tide line in the Chesapeake Bay watershed.[15] Development and changes to the natural environment in this area are subjected to rigorous oversight because their potential impact on the bay. The law also established the Critical Area Commission and charged it with creating a protection program to guide development in the critical area; this mandate gave rise to the Chesapeake Bay Critical Area Protection Program. While the law and protection program is not designed to halt development, they require local jurisdictions to work with the commission to establish standards for development to ensure their zoning ordinances, land use codes, and subdivision regulations support: "the protection of water quality, the conservation of habitat, and the accommodation of future growth and development without adverse impacts."[16]

In order to move forward with National Harbor, the team had to satisfy multiple levels of review related to the Chesapeake Bay Critical Area Protection Program and its oversight by the Critical Area Commission. The team also worked with the Maryland Department of the Environment to obtain a range of specialty permits, such as a Board of Public Works Tidal Wetlands License for work on the National Harbor piers and surrounding area. Various environmental agreements, including a Conservation Agreement with Prince George's County Department of Environmental Resources and a Chesapeake Bay Critical Area Conservation Agreement, were put in place to guide environmentally sensitive aspects of the development. The development plan also featured a dredging plan, the incorporation of trails and natural buffers around the property, tree conservation planning, and the construction of a new sanitary sewage pumping station to provide adequate sewer capacity for the project. Additionally, the Peterson Companies contributed $1 million to Maryland's tidal wetlands restoration fund. Further, environmental considerations at National Harbor have continued well after the necessary project approvals were obtained. Ongoing environmental efforts include contributing to the creation of new artificial reef structures to shelter marine life and encourage increases in the Potomac River bass population. While it has become a destination in its own right, the success of National Harbor will always be, in some ways, tied to preserving the natural environment that its residents and patrons enjoy along the Potomac River, a fact that has always been appreciated by the development team.

In addition to environmental challenges, the team also had to navigate sensitive cultural conditions. The land surrounding the National Harbor site was once part of the Oxon Hill Manor plantation belonging to Thomas Addison in the 1700s.[17] Part of the Oxon Hill Manor property was eventually sold to the Bayne family, who established the Salubria Plantation on their portion of the land. Over the centuries, these large land-holdings were eventually broken up and most of the historic buildings were destroyed. In 1978, a portion of the Oxon Hill Manor property, including a new manor house built in 1928, was listed on the National Register of Historic Places and eventually transferred to M-NCPPC.[18] The land associated with this historic property is adjacent to the southern portion of National Harbor's Beltway Parcel. Other portions of Oxon Hill Manor and the Salubria Plantation were transferred through various sales over the years and eventually became part of the land included in National Harbor's Beltway and Tanger Parcels.

Given the historic nature of the area, multiple archeological studies have been conducted on different portions of the National Harbor site, beginning as early as 1987 when the property was still part of the PortAmerica project.[19] The Maryland Historical Trust (MHT) required archeological reports as part of the plan of subdivision review and approval process and various Phases I, II, and III archeological reports were conducted between 2003 and 2012. These were compliant with Section 106 requirements of the National Historic Preservation Act of 1966 (NHPA)[20] and were submitted, as required, to the Prince George's County Historic Preservation Commission (HPC), which is tasked with protecting properties included on the county's inventory of historic sites.[21]

Of particular concern were the Addison Family Cemetery on the Beltway Parcel and the remains of artifacts from slave quarters that once existed on the part of the Salubria Plantation now part of the Tanger Parcel. Complicating matters, the cemetery, which had not been considered historic during permitting of the original PortAmerica project nor when National Harbor project began in the late 1990s, became a designated historic site in 2010.[22] Special considerations for the treatment of the cemetery were included both in the 1998 CSP and in a subsequent 2000 Memorandum of Agreement with the MHT[22]; nonetheless, its change to designated historic status created an unexpected challenge for the development team.

When working with historically and culturally sensitive sites, adopting a diligent, conscientious approach to the importance of the underlying and often irreplaceable heritage is the best way for a developer to navigate the challenges that can arise. Depending on specific circumstances, this may increase development costs due to preservation efforts, require a partial project redesign, and involve working with preservation groups and other stakeholders, along with a range of other approaches. In addition to working with the NHT and HPC, the National Harbor team also engaged with surviving ancestors of the Addison Family and the Prince George's County African American Heritage Preservation Group to find solutions that would allow development to continue while showing due respect for the historical events, places, and people associated with the site. As part of its efforts, the Peterson Companies retained and persevered artifacts recovered from the site during early archeological studies and from prior site work on PortAmerica; created educational outreach videos about the history of the area; and applied for Historic Work Permits. The team also endeavored to incorporate historical acknowledgments in the project design. Through efforts led by the African American Heritage Preservation Group, the Potomac River Heritage Visitors Center, and Salubria Memorial Garden were created as part of the Tanger Outlets. These dedicated venues display artifacts from the site and host the "Experience Salubria" exhibition, which aims to honor Prince George's County's rich cultural and African American heritage while educating visitors about the history and experience of enslaved people.[23] On the Beltway Parcel, a memorial marks the original Addison Family Cemetery, which was relocated to an historic church approved by the family so that the remains would not be disturbed by construction nor be surrounded by a high-density mixed-use development rather than the quiet rural environment that existed during their original interment.

CHAPTER CONCLUSION

National Harbor is a complex mixed-use project that has continued to evolve over a decades-long development period. A challenging yet alluring site along the Potomac River, it has always held the possibility of greatness. After the failed efforts of previous developers, Milt Peterson and his team were able to unlock the site's potential and achieve overwhelming success with their vision for National Harbor. Doing so required modifying zoning ordinances; weathering the Great Recession; capturing the synergy of an array of different uses; coordinating countless applications, permits, and construction teams; and navigating sensitive environmental and cultural conditions. Thanks to the perseverance of the Peterson Companies and the National Harbor team, today the project is the flagship location of Prince George's County, serving local residents, the greater Washington, D.C. area, and international visitors alike.

REFERENCES

1. U.S. Census Bureau website, 2017 American Community Survey 5-Year Estimates, available at https://factfinder.census.gov/faces/nav/jsf/pages/community_facts.xhtml?src=bkmk (accessed February 10, 2019).

2. Maryland Department of Transportation, State Highway Administration Office of Planning and preliminary Engineering, Data Services Division, "AADT's of Stations for the Years 2010–2016" report available at https://www.roads.maryland.gov/OPPEN/Station_history.pdf (accessed July 20, 2018); note Maryland Department of Transportation traffic count data show more than 191,000 cars per day in 2010 and more than 230,000 cars per day in 2016 near the I-495/Woodrow Wilson Bridge - I-295 interchange; the I-495 - US Route 210 interchange shows more than 140,000 cars per day in 2010 and more than 175,000 cars per day in 2016.

3. J. T. Lewis, *The Tragedy of PortAmerica: And Other Developments from Tysons Corner to Istanbul*, New York, New York: Vantage Press, Inc., 2012.

4. "PortAmerican Challenged" article in the Washington Post by Eugene L. Meyer, July 11, 1986, available at https://www.washingtonpost.com/archive/local/1986/07/11/portamerica-challenged/42a62811-9dae-446c-be37-a76592300ae1/?utm_term=.42c4dc2a9904 (accessed July 19, 2018).

5. Public Law 99-215 (99 STAT. 1724); J. T. Lewis, *The Tragedy of PortAmerica: And Other Developments from Tysons Corner to Istanbul*, New York, New York: Vantage Press, Inc., 2012.

6. Public Law 101-73, "Financial Institutions Reform, Recovery, and Enforcement Act of 1989," August 9, 1989; J. T. Lewis, *The Tragedy of PortAmerica: And Other Developments from Tysons Corner to Istanbul*, New York, New York: Vantage Press, Inc., 2012; "The RTC'S Rigidity Helped Make PortAmerica a Business Tragedy," article in the Washington Post by Rudolph A. Pyatt Jr., August 31, 1995, available at https://www.washingtonpost.com/archive/business/1995/08/31/the-rtcs-rigidity-helped-make-portamerica-a-business-tragedy/3d553c88-505a-471d-bdbd-70ea7793d520/?utm_term=.5b06923d643f (accessed July 19, 2018).

7. Tanger Outlets Annual Report 2013; "Prince George's County Executive Baker, the Peterson Companies Welcomed Tanger Outlets National Harbor at Groundbreaking" press release from Prince George's County Economic Development Corporation, available at http://www.pgcedc.com/press-release/149-prince-george-s-county-executive-baker-peterson-companies-welcomed-tanger-outlets-national-harbor-at-groundbreaking (accessed August 5, 2018).

8. "Destination DC Announces Record 20.8 Million Domestic Visitors to Washington, DC in 2017" press release by Destination DC, May 9, 2018, available at https://washington.org/press-release/destination-dc-announces-record-domestic-visitors-2017 (accessed August 5, 2018).

9. MGM Resorts International Annual Report 2014, pg 59 Note 11 to the accompanying consolidated financial statements, available at https://s22.q4cdn.com/513010314/files/doc_financials/annual/2014/2014-252BAnnual-252BReport.MGMResortsInternational.pdf (accessed August 4, 2018).

10. "MGM gets Prince George's council backing for casino construction at National Harbor" Washington Post article by Luz Lazo, July 21, 2014, available at https://www.washingtonpost.com/local/trafficandcommuting/mgm-gets-prince-georges-council-backing-for-casino-construction-at-national-harbor/2014/07/21/c4804bb4-10e2-11e4-9285-4243a40ddc97_story.html?utm_term=.6fecb5f36423 (accessed July 20, 2018).

11. USGBC website, MGM National Harbor project overview available at https://www.usgbc.org/projects/MGM-National-Harbor (accessed August 19, 2018).

12. Marriott International press release "Marriott International Expands Group and Meetings Portfolio with Acquisition," May 31, 2012, Bethesda, Maryland, available at Marriot International Investor Relations website at https://marriott.gcs-web.com/news-releases/news-release-details/marriott-international-expands-group-and-meetings-portfolio (accessed August 4, 2018).

13. Public Law 106–53 (113 STAT. 385), Section 597: National Harbor Maryland, August 17, 1999.

14. Prince George's County Council website, section "Conceptual & Detailed Site Plan Review" under section "Planning & Development Process" available at https://pgccouncil.us/215/Conceptual-Detailed-Site-Plan-Review (accessed August 12, 2018).

15. Maryland Natural Resources Code Ann. § 8-1807; Maryland Department of Natural Resources website, Critical Area Commission General Information link, section on "Background and History" available at http://dnr.maryland.gov/criticalarea/Pages/background.aspx (accessed August 18, 2018).

16. "Bay Smart. A Citizen's Guide to Maryland's Critical Area Program," Critical Area Commission for the Chesapeake and Atlantic Coastal Bays, Annapolis, Maryland, printed September, 2007 Revised December, 2008, edited by Mary R. Owens, DNR Publication Number 10-9182007-248, available for download at https://dnr.maryland.gov/criticalarea/Documents/baysmart.pdf (accessed August 18, 2018).

17. Maryland Historical Trust Phase II and Phase III Archeological Database and Inventory report for site number 18PR692; Maryland National Capital Park and Planning Commission memo regarding HAWP 2016-036 to the Historic Preservation Commission of February 13, 2017.

18. U.S. Department of the Interior, National Park Service National Register of Historic Places website digital archive of National Register Information System ID78003117, available at https://npgallery.nps.gov/NRHP/AssetDetail?assetID=e37762cc-2606-4020-a380-5bd5a1bf2a79 (accessed September 2, 2018); Maryland National Capital Parks and Planning Commission website, section on Sites & Museums—Oxen Hill, available at http://www.pgparks.com/3019/Plan-Your-Visit (accessed September 2, 2018).

19. Maryland Historical Trust Phase II and Phase III Archeological Database and Inventory report for site number 18PR370.

20. Advisory Council on Historic Preservation website, section on An Introduction to Section 106, available at https://www.achp.gov/protecting-historic-properties/section-106-process/introduction-section-106 (accessed September 1, 2018); Maryland Historical Trust Phase II and Phase III Archeological Database and Inventory report for site number 18PR692.

21. Historic and Architectural Significance: Evaluation Criteria, Policies and Guidelines, Prince George's County Historic Preservation Commission, available at http://www.mncppc.org/DocumentCenter/View/615/HPC-Evaluation-Criteria-PDF (accessed September 1, 2018).

22. Maryland National Capital Park and Planning Commission memo regarding HAWP 2016-036 to the Historic Preservation Commission of February 13, 2017.

23. "Using painful pieces of history, Prince George's hopes to boost tourism," article in the Washington Post by DeNeen L. Brown, June 3, 2014, available at https://www.washingtonpost.com/local/using-painful-pieces-of-history-prince-georges-hopes-to-boost-tourism/2014/06/03/525873c4-eb51-11e3-9f5c-9075d5508f0a_story.html?utm_term=.021c7f5ea64e (accessed September 3, 2018).

INDEX